NIKOLA TESLA
AND
POLYPHASE POWER

NIKOLA TESLA
AND
POLYPHASE POWER

By

Eric P. Dollard

PUBLISHED BY
EMEDIAPRESS.COM
SPOKANE, WASHINGTON

Cover Layout: Aaron Murakami
Transcribed by: Griffin G. Brock
Edited by: Griffin G. Brock, Simon Davies

Digital Version   Released: November 2023
Print Version   Released: November 2023

ISBN-13: 9798864478448

Digital Edition Published by: Emediapress.com
                                  PO Box 10029,
                                  Spokane, WA 99209
                                  https://emediapress.com

EPD Laboratories

EPD Laboratories, Inc.
PO Box 10029
Spokane, WA 99209

info@epdlabs.org
https://ericpdollard.com

Support EPD Laboratories, Inc. 501(c)3 non-profit organization
with tax-deductible donations at https://ericpdollard.com

Table Of Contents

The Engineering Achievements of Nikola Tesla

The System of Polyphase Power, which is the subject of the following text, was conceived in the latter part of the 19th Century by Nikola Tesla (1888).

Nikola Tesla, 1856 to 1943, may be considered the most enigmatic character in recent history, adored by some, despised by others. However, it cannot be denied that his Polyphase System, in collaboration with the efforts of George Westinghouse, transformed America, and ushered in the Electric Age.

While it is generally known "who was" Nikola Tesla, it is not so well known "what was" Nikola Tesla.

Nikola Tesla acquired a classical education in the universities of 19th Century Europe. Here his efforts were directed into the study of Electrical Engineering, his interest starting at a young age.

This was a relatively new field of study since the practical application of Electric Forces was just beginning at this point in history, that is, the Electric Telegraph and the Electric Light. Hence, based upon his interests and education, Tesla was to become an Electrical Engineer.

Tesla acquired considerable practical experience working as an electrical engineer in the field of Electric Light and Power. Here he recognized the need for a superior Electric Motor, and thus began his efforts to create such a machine. Out of this effort came his conception of the Rotating Magnetic Field. Ultimately, his long and arduous effort made him quite wealthy, for a while.

With his newly acquired wealth and resources, Nikola Tesla established a new laboratory. Here he undertook his next major effort, High Voltage, High Frequency Phenomena. Many of Nikola Tesla's most profound, and least understood, discoveries were made in this laboratory.

So stunning were his extraordinary displays of high voltage effects, in addition to the remarkable impact his experimental researches had on the electrical science of that era, Nikola Tesla would achieve world fame, for a while. Here would be the scenario for his next major achievement, the Longitudinal Electric Field.

One of Nikola Tesla's principal efforts in this laboratory was the study of Electric Rays. He was continuing the work of Sir William Crookes, a leading pioneer in that field of endeavor, and someone who Tesla held in high regard. This was the era of the great minds in Electric Ray research, Lenard, Roentgen, and J.J. Thomson. Tesla had now become an electrical scientist, or what then was called an "Electrician".

Tesla would have a distinct advantage over his contemporaries, for while everyone struggled to obtain effects with feeble and primitive laboratory apparatus, Tesla was to obtain very powerful effects due to the refinement of his apparatus. Remarkable was his ability to apply his engineering knowhow to his scientific endeavors. The apparatus of others would only produce milliwatts while the apparatus of Tesla would produce kilowatts. Accordingly, the Electric ray research of Tesla excelled over his contemporaries, for a while.

In the course of his laboratory work, Tesla observed that his apparatus created a Longitudinal Electric Field. He knew that Herman von Helmholtz had theorized the existence of Longitudinal Electric Waves. Accordingly, Tesla presented his findings to Helmholtz, who in turn verified Tesla's expectations.

Nikola Tesla's expectations were the transmission and reception of Electric Forces without the use of connecting wires. In a very roundabout way this would eventually lead to what has become commonly known as "Radio". However, the Electric Waves employed by Tesla were not the Electric Waves to be employed by contemporary Radio. Here begins the enigma of Nikola Tesla.

The 20th Century would have no use for Nikola Tesla, nor his ideas of Longitudinal Electric Waves. Modernistic science declared that such waves cannot possibly exist.

Thus ends the story of a scientist and engineer, Nikola Tesla.

Goethe and the Vision

Throughout much of Nikola Tesla's life he would be plagued by severe hallucinogenic episodes. Despite this condition Tesla would learn to master it and put it to work for him. Accordingly, this provided him with an enhanced power of visualization.

The mind of Nikola Tesla had been overly occupied with the ideas relating to A.C. rotating machinery, most notably, the A.C. motor, the lacking element in electric utility system of his day.

Historically, Nikola Tesla would claim his revelation of the Rotating Magnetic Field, and its Polyphase Power System, that came to him in the form of an all-encompassing vision, or by his account, as a "flash of light".

As the Sun was setting Tesla was reciting from memory certain verses written by Johann Goethe;

> *The glow retreats, done is the day of toil;*
> *It yonder hastes, new fields of life exploring;*
> *Ah, that no wing can lift me from the soil;*
> *Upon its track to follow, follow soaring.*

At this moment the vision in Nikola Tesla's mind was the Sun acting as a whirling magnetic pole, carrying him around the Earth, to follow soaring. Hence the Rotating Magnetic Field was born.

The moment the Sun was setting at Nikola Tesla's location, it was, at the same moment, at its noon position one quarter turn around the Earth ahead of him, figure (1). At this same moment, the Sun was at its sunrise position two quarter turns around the Earth ahead of him, figure (2). Likewise, at this same moment in time, the Sun was at its midnight position three quarter turns around the Earth, figure (3). Finally, at the same moment in time, four quarter turns around the Earth, is the location of Nikola Tesla at sunset. The cycle is complete, consisting of "Four Phases", this in the mind of Tesla, figures (1, 2, 3, 4, and 5).

Hereby, the four-unit positions of the Sun are:

1. Midnight (MN), start position
2. Sunset (SS), first position
3. Noon (ON) second position
4. Sunrise (SR) third position

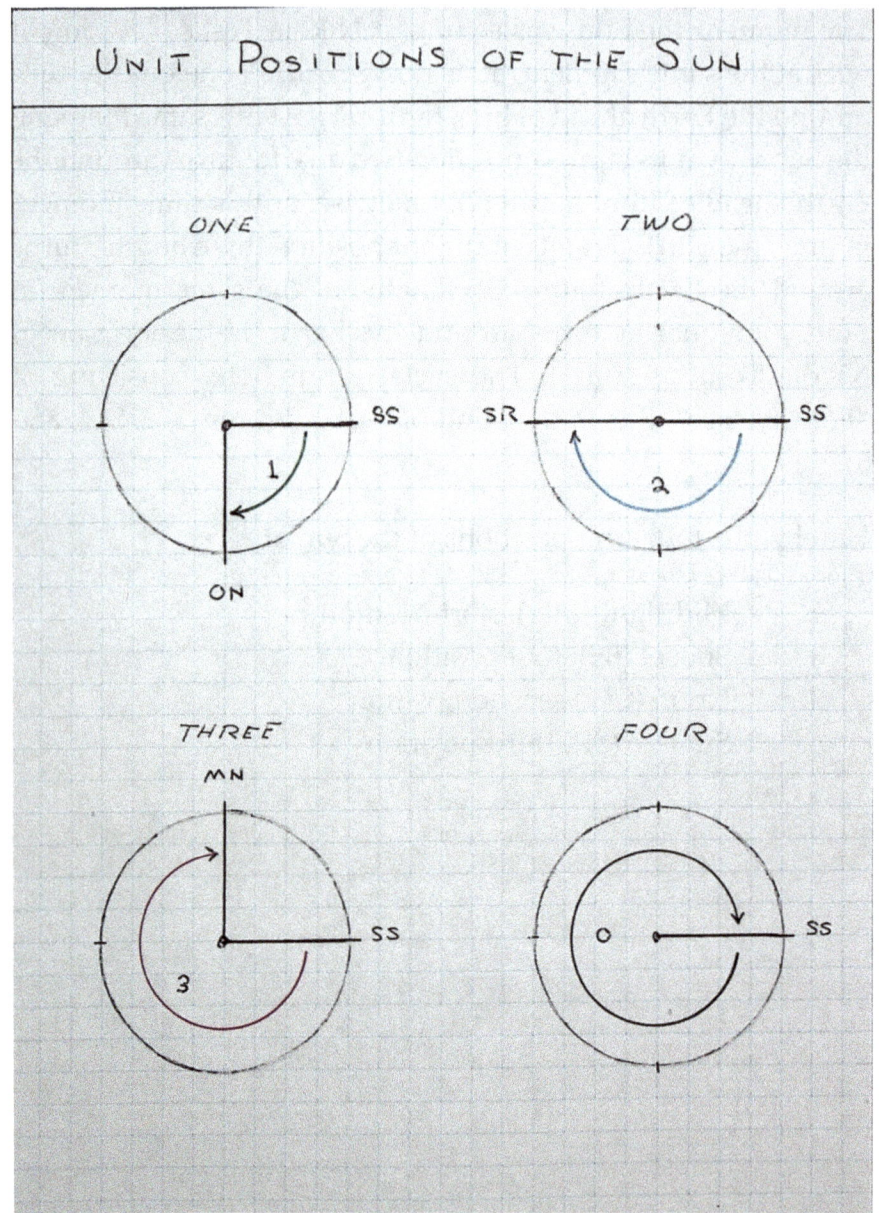

Figs. 1, 2, 3, & 4 – Unit positions of the Sun

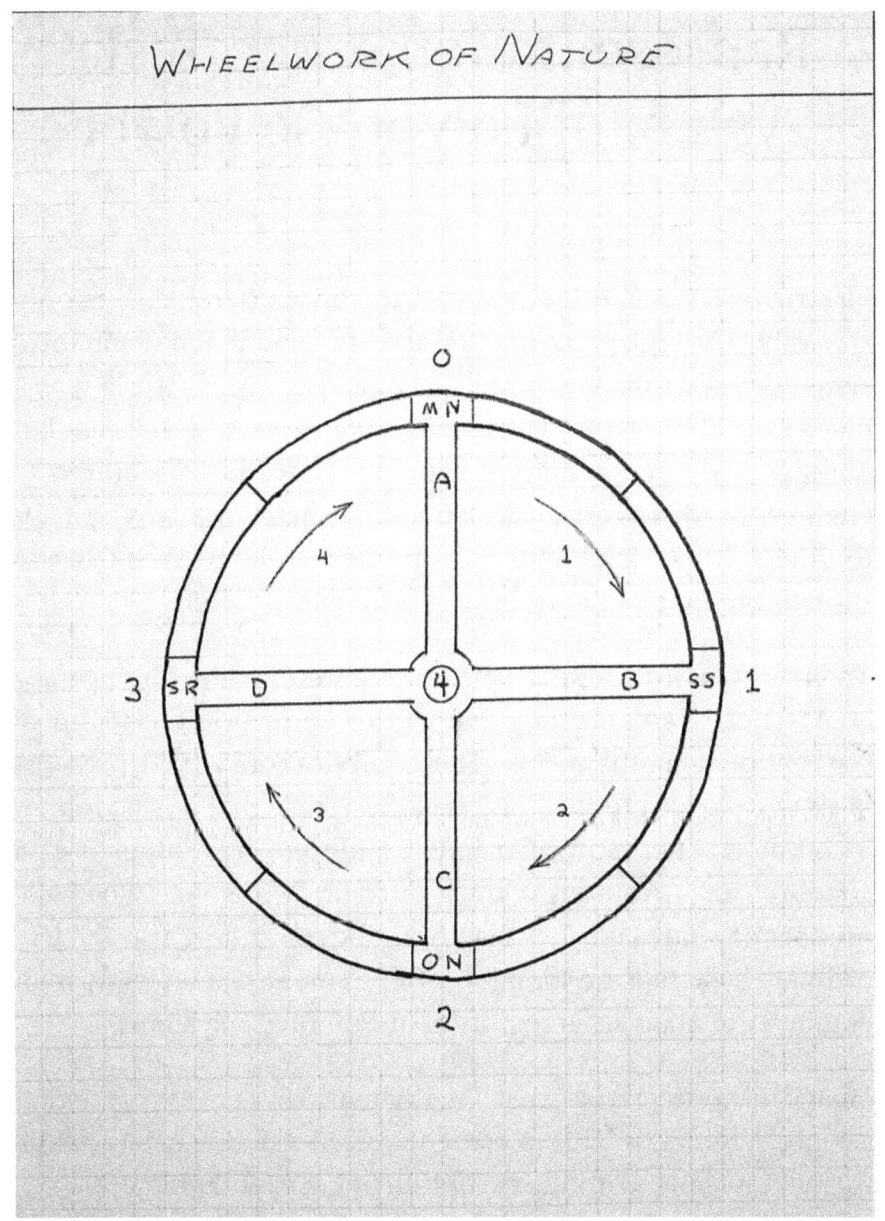

Fig. 5 – The wheelwork of nature

Representation of the Tesla System of Polyphase Power

Electricity resides in the realm of the imponderable. It is only accessible through experimental research and the mathematical symbolism that these researches engender. Nikola Tesla was a master in the realm of experimental research, but he left nothing of the mathematical symbolic procedures required to engineer his discoveries. He stated that such work was best left to others. Most notable of the so-called "others" was Charles Proteus Steinmetz, the wizard of Schenectady. Ref. [1]

The mathematical representation required for the engineering analysis of Nikola Tesla's discoveries is outside the realm of conventional mathematics. The complication which presents itself with regard to the Polyphase System is that it is no longer a common Alternating Current Condition, but rather it is a condition of a Rotating Direct Current. Moreover, when the number of phases is other than four, the common Cartesian Co-ordinate System is no longer useful. Evidently, a novel form of engineering analysis is required for Polyphase Systems.

Heaviside once wrote that "mathematics is an experimental science", Ref. [2]. This comment raised the ire of academic mathematicians, with the exception of Charles Steinmetz.

Charles Steinmetz developed a revolutionary system of mathematics very well suited for the analysis of Alternating Current Systems. This he would call the "Symbolic Method". Ref. [3]

8

In his time, this would become known as the "Steinmetz Method", Ref. [4]. In 1897 he published his first engineering textbook titled "Theory and Calculation of Alternating Current Phenomena", which demonstrated the application of the Symbolic Method to the solution of the A.C. engineering problems encountered in practice. The publication of this textbook, and its widespread acceptance would elevate Steinmetz to world fame, Ref. [5], the one thing he held in common with Nikola Tesla.

The material presented in his book would also provide the groundwork for the analysis of Polyphase Systems, Ref. [6], in particular, his so-called "Roots of The Unit". Ref. [6a] This is the basis for a more advanced Polyphase Theory. The work of Charles Steinmetz in this regard received further development by his protégé, Ernst Alexanderson, Ref. [7], and of this arrived the final established form of Polyphase Analysis, that of Charles Fortescue. Ref. [8]

Fortescue extended the work of Steinmetz and Alexanderson into what may be defined as the "Method of Multiple Co-ordinate Systems". This established a solid mathematical basis for the analysis of Polyphase Systems. Moreover, this development would advance the General Theory of Numbers, and mathematics in general. Ref. [9] Today the "Fortescue Method" is known as the method of "Symmetrical Components". Ref. [10] In the words of Fortescue, it is called a "Sequence Algebra". Ref. [11]

Four Phase Mathematics

The mathematical analysis of any Polyphase System is founded upon that methodology developed by Charles Fortescue. This method of analysis will be developed throughout the following articles of this text.

The "Fortescue Method" employs a set of Multiple Co-ordinate Systems, one for each phase. Hereby, the Tesla Four Phase System is analytically represented by four specific Co-ordinate Systems, one for each of the four phases, A, B, C, and D. These are portrayed by figure (6).

Each Co-ordinate System consists of four co-ordinates, which have been established as;

Midnight, 0   Sunrise, 1
Noon, 2   Sunset, 3

Specific arrangements of these co-ordinates present themselves in each phase, as shown in figure (6). Each specific co-ordinate arrangement is identified as a specific "Sequence". The interaction of these sequences is essential to the working of the Fortescue Method. Accordingly, Fortescue refers to his method in terms of "Sequence Algebra".

In the analysis of Polyphase Systems, several orders of co-ordinate sequences present themselves. The number of these sequences is directly related to the number of phases.

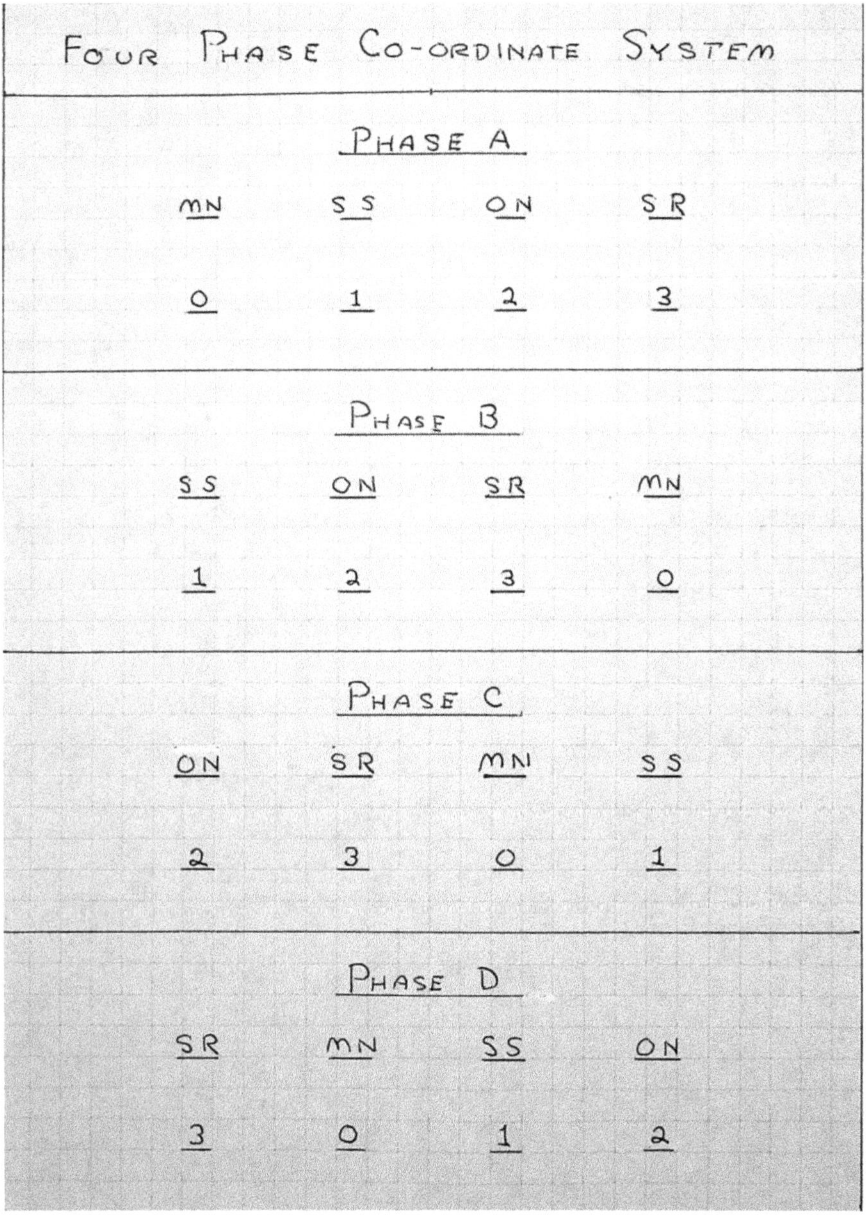

FOUR PHASE CO-ORDINATE SYSTEM			
PHASE A			
MN	SS	ON	SR
0	1	2	3
PHASE B			
SS	ON	SR	MN
1	2	3	0
PHASE C			
ON	SR	MN	SS
2	3	0	1
PHASE D			
SR	MN	SS	ON
3	0	1	2

Fig. 6 – Co-ordinate system of Four Phase

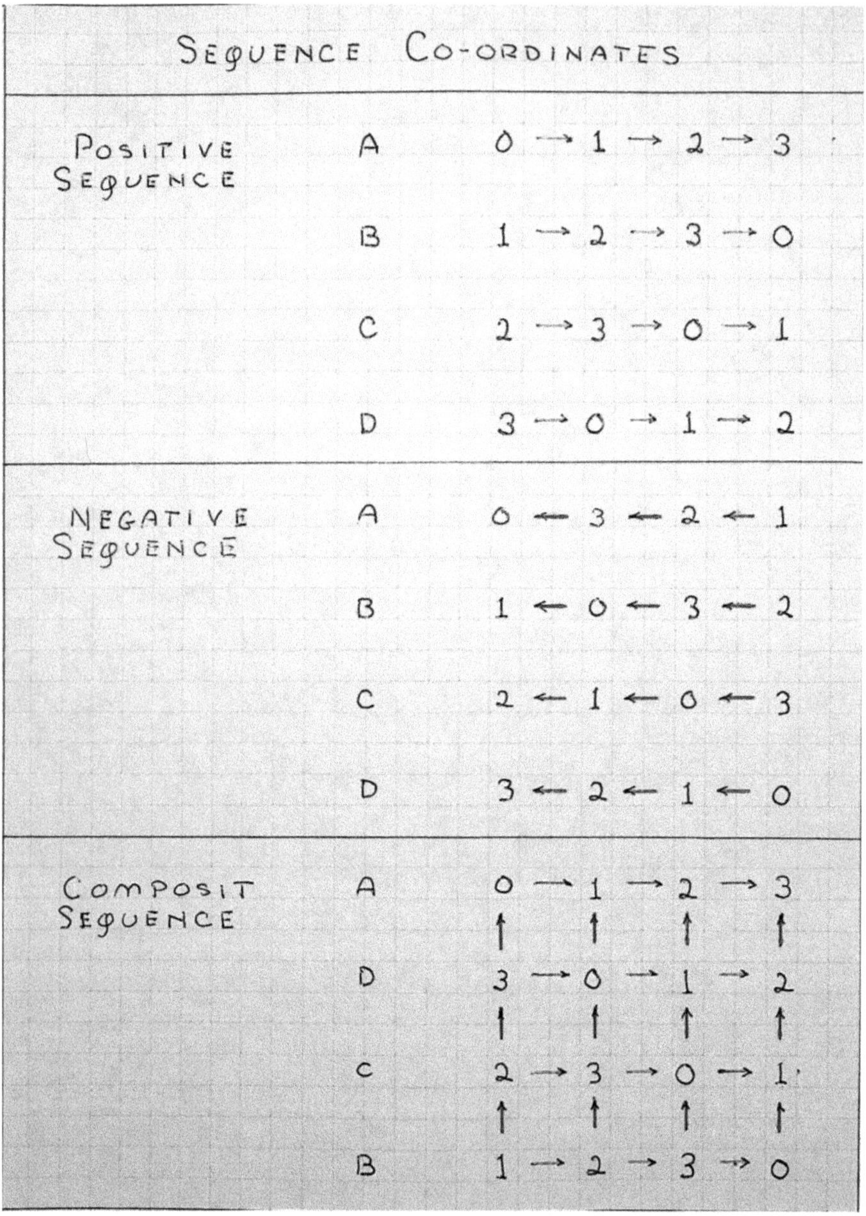

SEQUENCE CO-ORDINATES

POSITIVE SEQUENCE	A	$0 \rightarrow$	$1 \rightarrow$	$2 \rightarrow$	3
	B	$1 \rightarrow$	$2 \rightarrow$	$3 \rightarrow$	0
	C	$2 \rightarrow$	$3 \rightarrow$	$0 \rightarrow$	1
	D	$3 \rightarrow$	$0 \rightarrow$	$1 \rightarrow$	2
NEGATIVE SEQUENCE	A	$0 \leftarrow$	$3 \leftarrow$	$2 \leftarrow$	1
	B	$1 \leftarrow$	$0 \leftarrow$	$3 \leftarrow$	2
	C	$2 \leftarrow$	$1 \leftarrow$	$0 \leftarrow$	3
	D	$3 \leftarrow$	$2 \leftarrow$	$1 \leftarrow$	0
COMPOSIT SEQUENCE	A	$0 \rightarrow$	$1 \rightarrow$	$2 \rightarrow$	3
	D	$3 \rightarrow$	$0 \rightarrow$	$1 \rightarrow$	2
	C	$2 \rightarrow$	$3 \rightarrow$	$0 \rightarrow$	1
	B	$1 \rightarrow$	$2 \rightarrow$	$3 \rightarrow$	0

Fig. 7 – Sequence of co-ordinates

Foremost are the Positive Sequence, and the Negative Sequence, the two most fundamental sequences of any Polyphase System. For a Four Phase condition these sequences are portrayed by figure (7).

The interaction of the Positive and Negative Sequences will engender other less obvious sequences. This is an objective in the application of the Fortescue Method. These other sequences can be arrived at by forming a Composite Sequence pattern. For the Four Phase condition, this is portrayed in figure (7).

In the Four Phase composite configuration shown, the Positive Sequence is positioned horizontal, and the Negative Sequence is positioned vertical. This results in a four-by-four square, or lattice, of Four Phase Co-ordinates.

Repositioning the Four Phase Co-ordinate lattice of figure (7) into the form of a Pythagorean Co-ordinate System, or "Lambdoma", figure (8), serves to emphasize the lesser order co-ordinate sequences. Two principle co-ordinate sequences present themselves as alternate horizontal rows, that is, the Even Alternate Sequence, and the Odd Alternate Sequence.

Unlike the Positive and Negative Sequences, the Alternate Sequences do not progress, but rather alternate, and thus represent an Alternating Current, or "Single Phase" condition.

In addition to the aforementioned Even and Odd Alternate Sequences, a third set of co-ordinates present themselves in the lattice of figure (8), this as a vertical row of zeros. This particular set of co-ordinates has no definite sequence and accordingly is called a Zero Sequence.

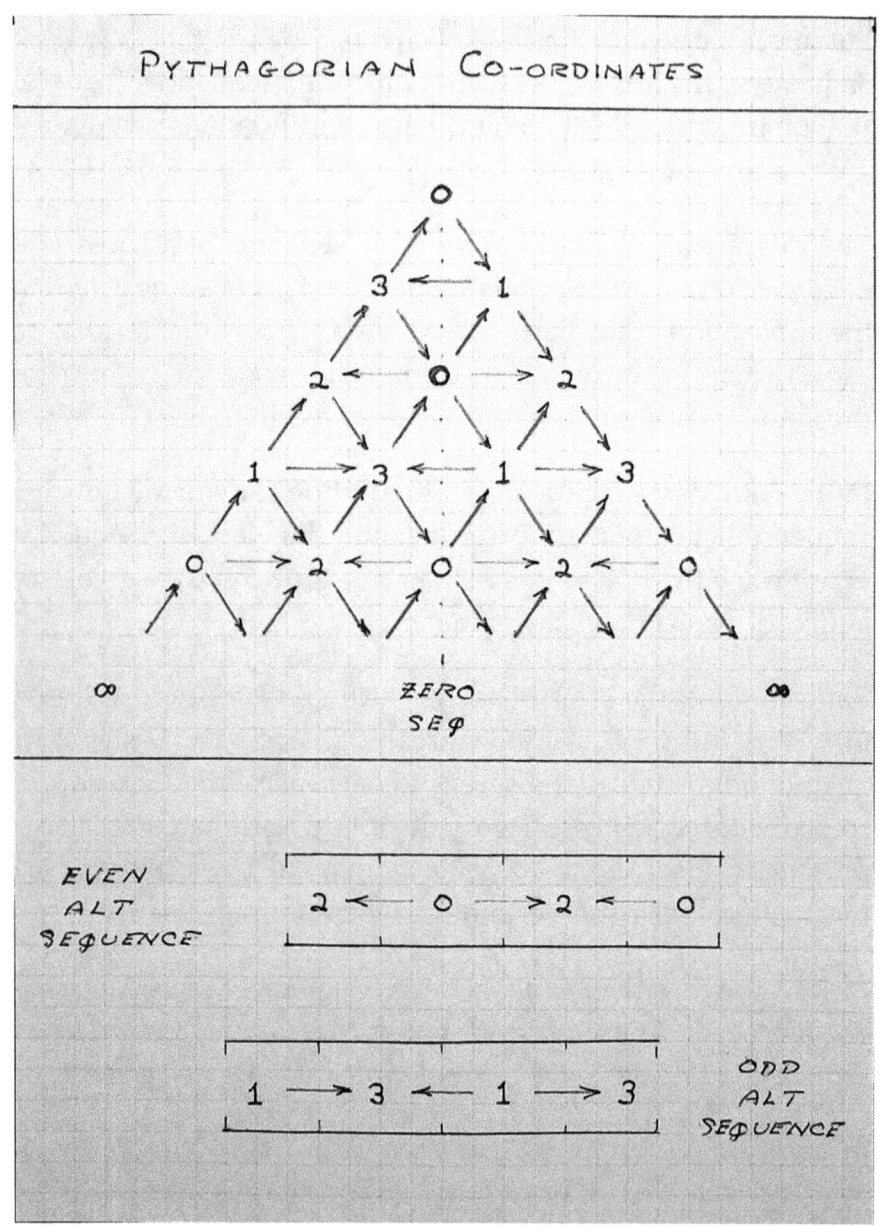

Fig. 8 – Pythagorean co-ordinates

In the development of Polyphase Theory, the analytical expression of the Zero Sequence component was the most difficult to arrive at. Ref. [12] Moreover, in any Polyphase Power System, the Zero Sequence is a parasitic mode best avoided or suppressed whenever possible.

For its existence, the Zero Sequence component required that the number of electric conductors in a Polyphase System exceeds the number of phases in that system by one. Thus, in a Four Phase System, a fifth conductor is required to support the Zero Sequence mode, and this additional conductor is called the "Neutral Conductor".

As portrayed by figure (9), the Four Phase condition yields a plurality of co-ordinate sequences, that is;

1.   The Positive and Negative pair of co-ordinate sequences which represent the rotational components of the Four Phase System, Positive for clockwise rotation, and Negative for counter clockwise rotation.

2.   The Even and Odd pair of co-ordinate Sequences which represent the Alternating components of the Four Phase System, Even for the co-sinusoidal alternation, and Odd for the sinusoidal alternation.

3.   The Zero, or Null, co-ordinate sequence which neither rotates nor alternates, but rather it defines a harmonic relationship within the Polyphase System.

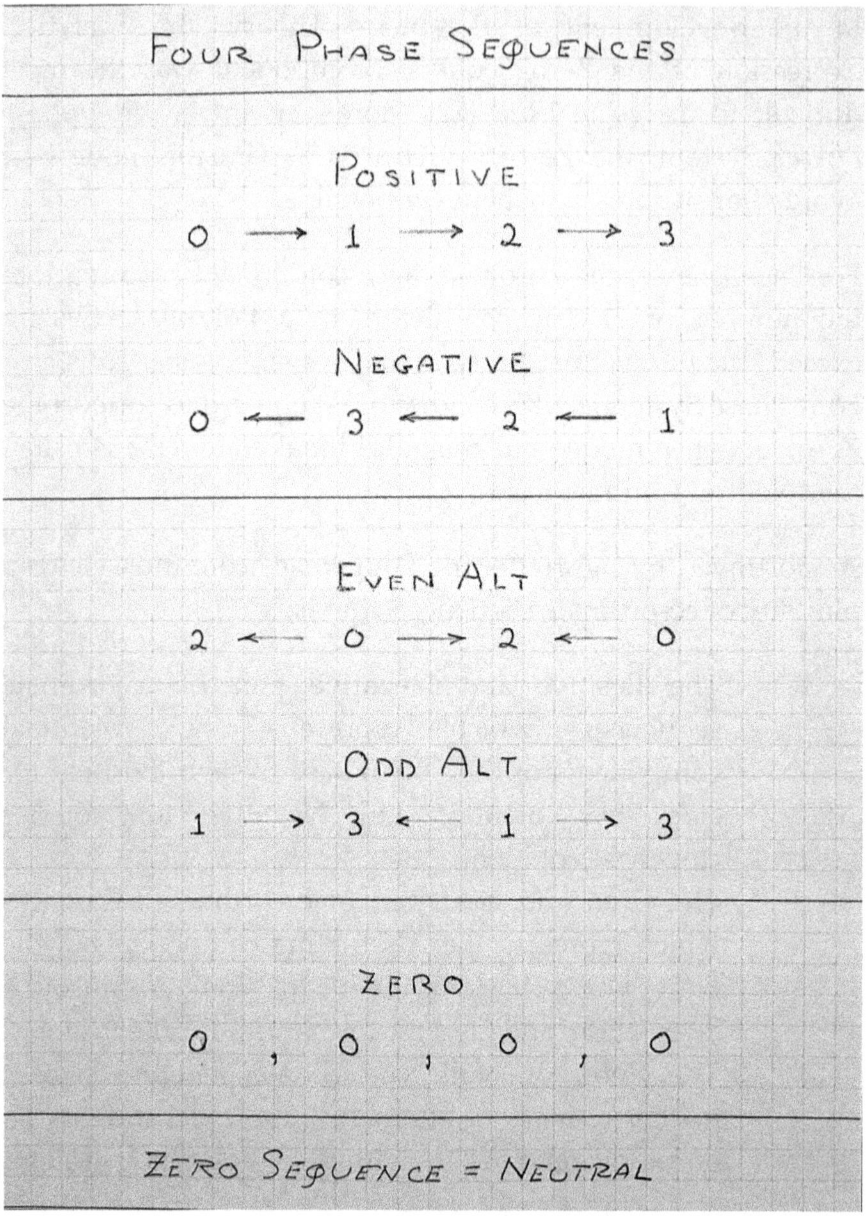

FOUR PHASE SEQUENCES

POSITIVE

0 —→ 1 —→ 2 —→ 3

NEGATIVE

0 ←— 3 ←— 2 ←— 1

EVEN ALT

2 ←— 0 —→ 2 ←— 0

ODD ALT

1 —→ 3 ←— 1 —→ 3

ZERO

0 , 0 , 0 , 0

ZERO SEQUENCE = NEUTRAL

Fig. 9 – Four Phase sequences

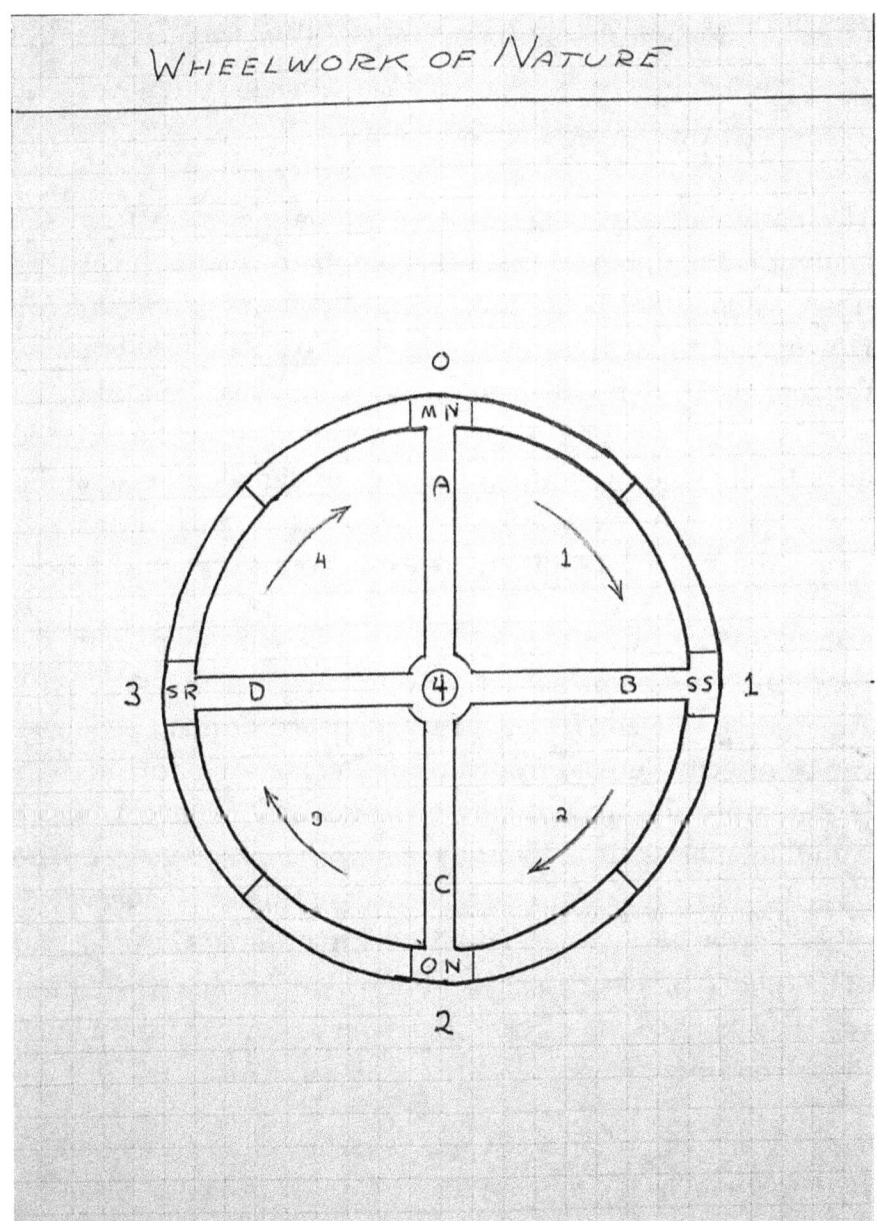

Fig. 5 – The wheelwork of nature

From an analytical standpoint, as well as from a practical standpoint, the Tesla Four Phase System offers several advantages over other numbered systems.

The particular advantage of a Four Phase System from the analytical standpoint is that this Polyphase condition is based upon a Quadrant-Polar arch form as shown by figure (5). However, this arch form is also the basis for Alternating Current itself. Ref. [13] Hence, the Four Phase condition, as with the Alternating Current condition, can be treated analytically in terms of the commonly understood Rectangular or Cartesian, Co-ordinate System, whereas the analysis of other numbered systems require more exotic forms of mathematics.

The particular advantage of a Four Phase System from the practical standpoint is its co-operative relationship with Alternating Current. Hereby, a rotational co-ordinate sequence can be directly derived from a complimentary pair of alternate (non-rotational) co-ordinate sequences. Conversely, a complimentary pair of alternate co-ordinate sequences can be directly derived from a rotational co-ordinate sequence. In other words, a Poly-Phase condition can be derived from a pair of so-called "Single Phase" conditions, or conversely, a pair of so-called "Single-Phase" conditions can be derived from a Poly-Phase condition.

This is an important advantage when transformation is required between "Single-Phase" and Poly-Phase Systems. Hence, this allows for unique interactions between the four phases and the four aspects of the Alternating Currents which constitute the Four Phase System.

The synthesis of a Rotating Sequence, either positive or negative, from a pair of complimentary Alternate Sequences, Odd or Even, is portrayed by figures (10) and (11). It is evident from figure (10) that by reversing a particular Alternate Sequence, Odd or Even, the result is the reversal of the Rotating Sequence. In figure (10), the reversal of the Odd Alternate Sequence reverses the Rotating Sequence from positive to negative.

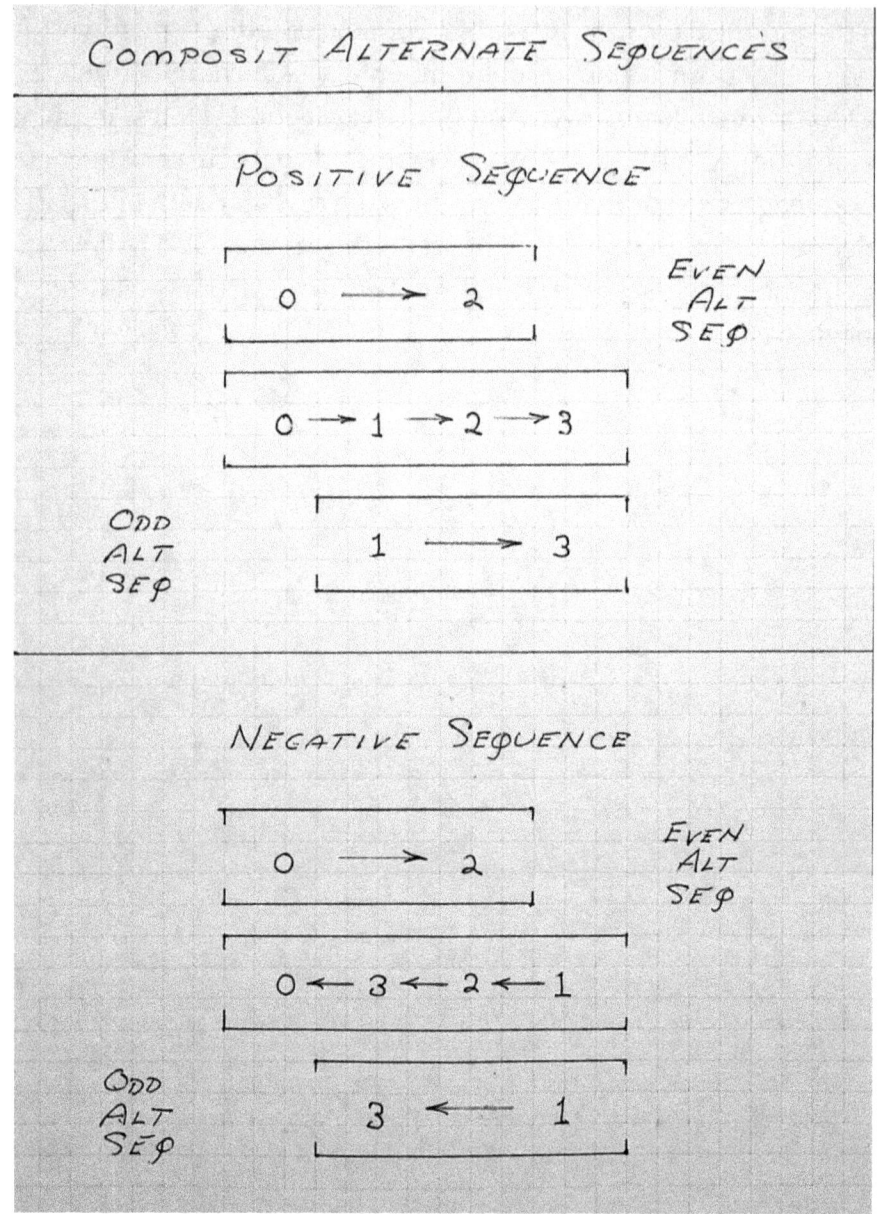

Fig. 10 – Composite Alternate Sequences

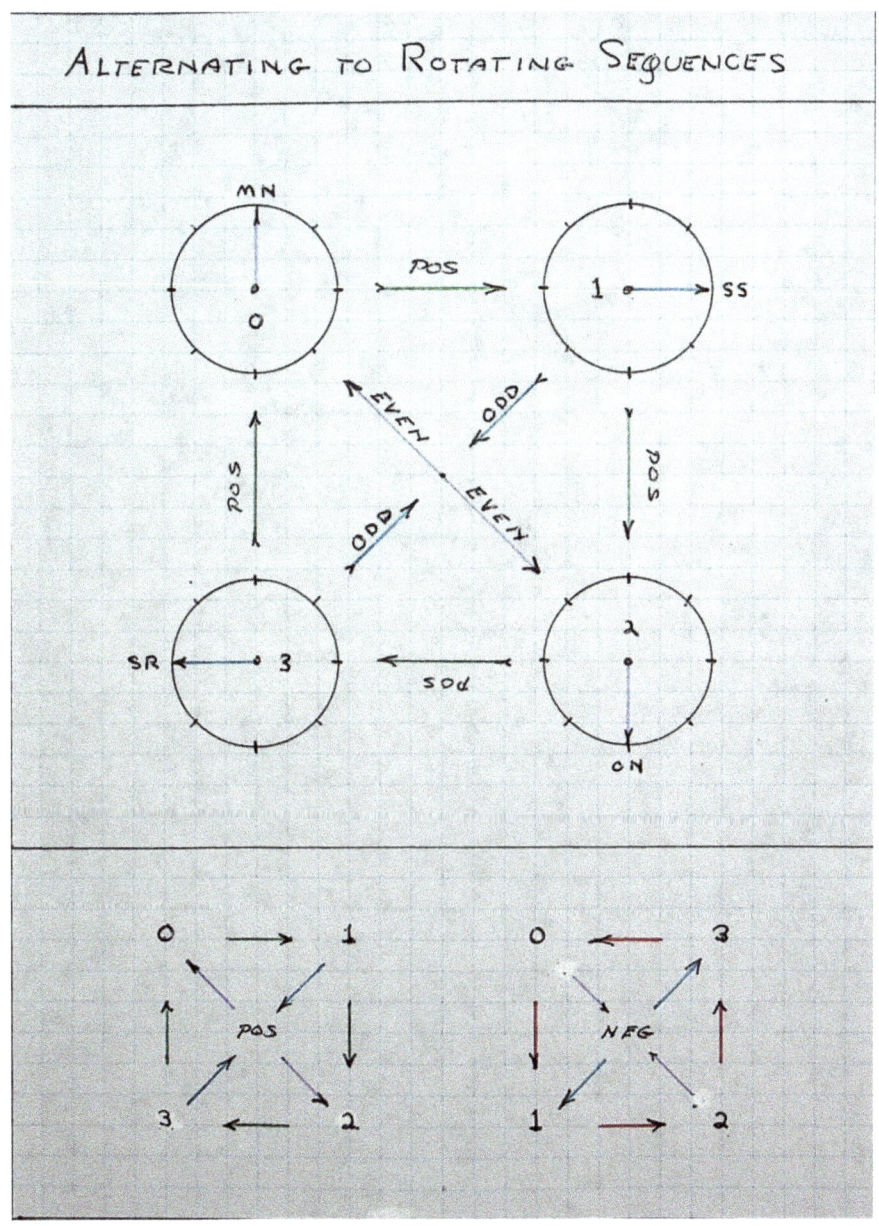

Fig. 11 – Alternating to Rotating Sequences

Fig. 12 – 1893 Chicago World's Fair
(The World's Columbian Exposition)

Fig. 13 – 1893 World's Fair "Electric Hall"

The Westinghouse-Tesla System of Electric Power

The first full scale application of the Tesla Four Phase System was constructed by the Westinghouse Electric Company for the purpose of providing light and power for the 1893 Chicago World's Fair. This was the largest such plant of that era, figure (12) and (13). Ref. [14]

Several sets of paired Synchronous Alternators of a rotating field construction were employed in this effort. The stator poles of one Alternator of the pair are displaced by one unit Four Phase step, or 90 degrees with respect to the stator poles of the other Alternator of the pair, figure (14) and (14a). Hereby this pair of Alternators are unified into a single Polyphase machine.

The rotating field coils of this pair of Alternators are configured so as to allow independent excitation for each sequence, Odd Alternate, Even Alternate, and Rotational, figure (15). Ref. [15]

At this point in history, it became common practice to designate each Alternate Sequence as a "Single Phase", and correspondingly designate the joint pair of Alternators as "Two Phase". Unfortunately, this practice exists to the present day. This stands in contradiction in the process of analysis.

By definition, the number of phases in any Polyphase System is numerically equivalent to the number of conductors required to produce a Rotation Sequence. Hence, a system of three phases requires three conductors, whereas a system of four phases requires four conductors, etc. Accordingly, the two conductors for an Alternate Sequence define a system of two phases, which is not a Poly-Phase System. Hereby, the pair of Alternators derives a pair of two conductor system, that is, four conductors which hence derives a "Four Phase System", figure (16). This figure presents a schematic portrayal of the quadrature arrangement of a pair of alternators which establish a Four Phase condition.

In common practice the mutual sets of armature coils (0, 1, 2, 3) are designated the "phases" of the system, and the external leading conductors (A, B, C, D) are designated the "lines" of the system. The phases are sometimes referred to as the "legs" of the system. Hence, a distinction is established regarding what relates to the phases, such as the "Phase Voltage", or what relates to the lines, such as the "Line Voltage". The relationship between phase and line values is dependent upon the particular configuration of any given Polyphase System.

A single unit of the World's Fair Power Plant is portrayed by figure (17) and (17a). The function of this plant is to convert the steam engine shaft horsepower into the electric line kilowatt power, this delivered to the electrical load.

In this setup the alternators, through the actions of the rotating field windings, figures (18), (18a), and (18b), serve to automatically regulate the transfer of the mechanical power of the engine to the electrical power of the load. The function of the alternators in this transfer of power is analogous to the function of the automatic transmission in an engine driven automobile.

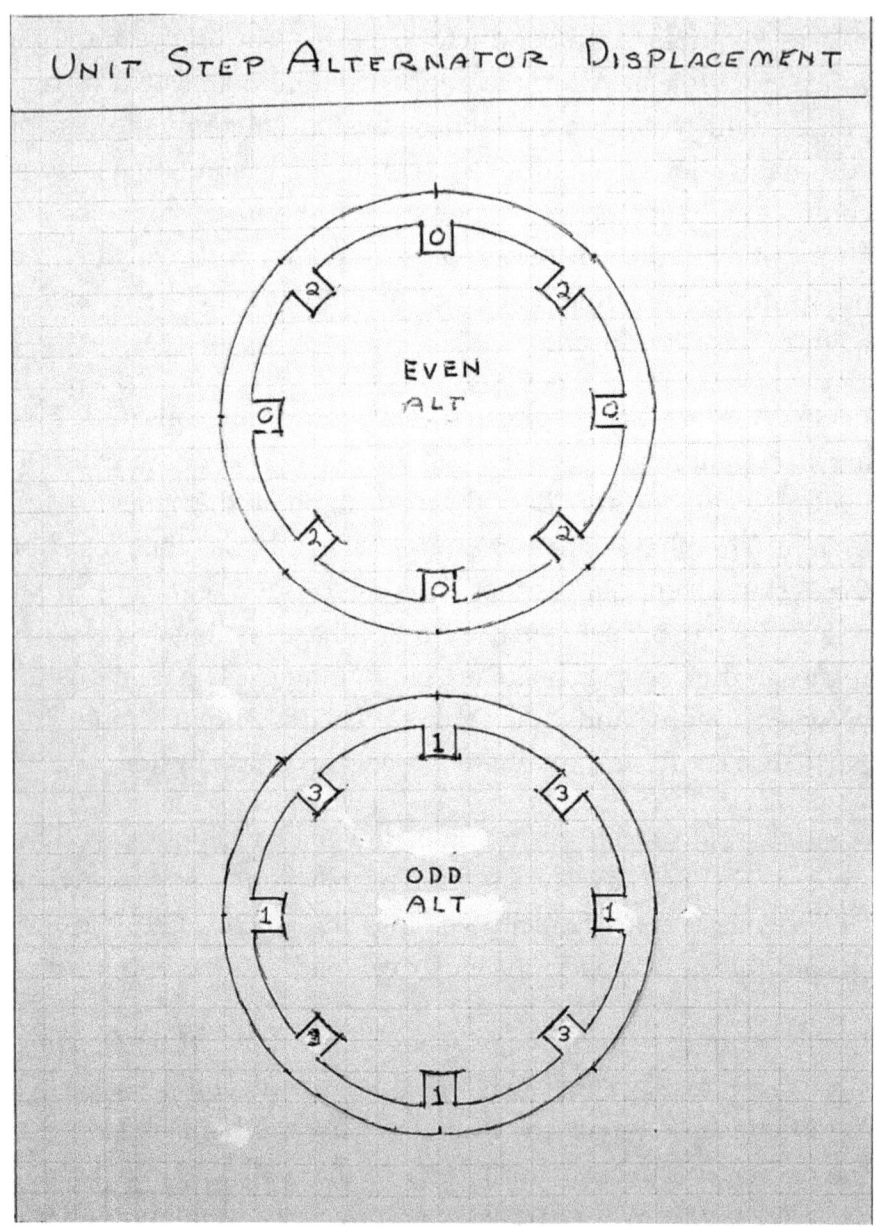

Fig. 14 – Alternator displacement unit step

Fig. 14a – Alternator stator winding assembly

Fig. 15 – Pair of alternators, showing slip rings

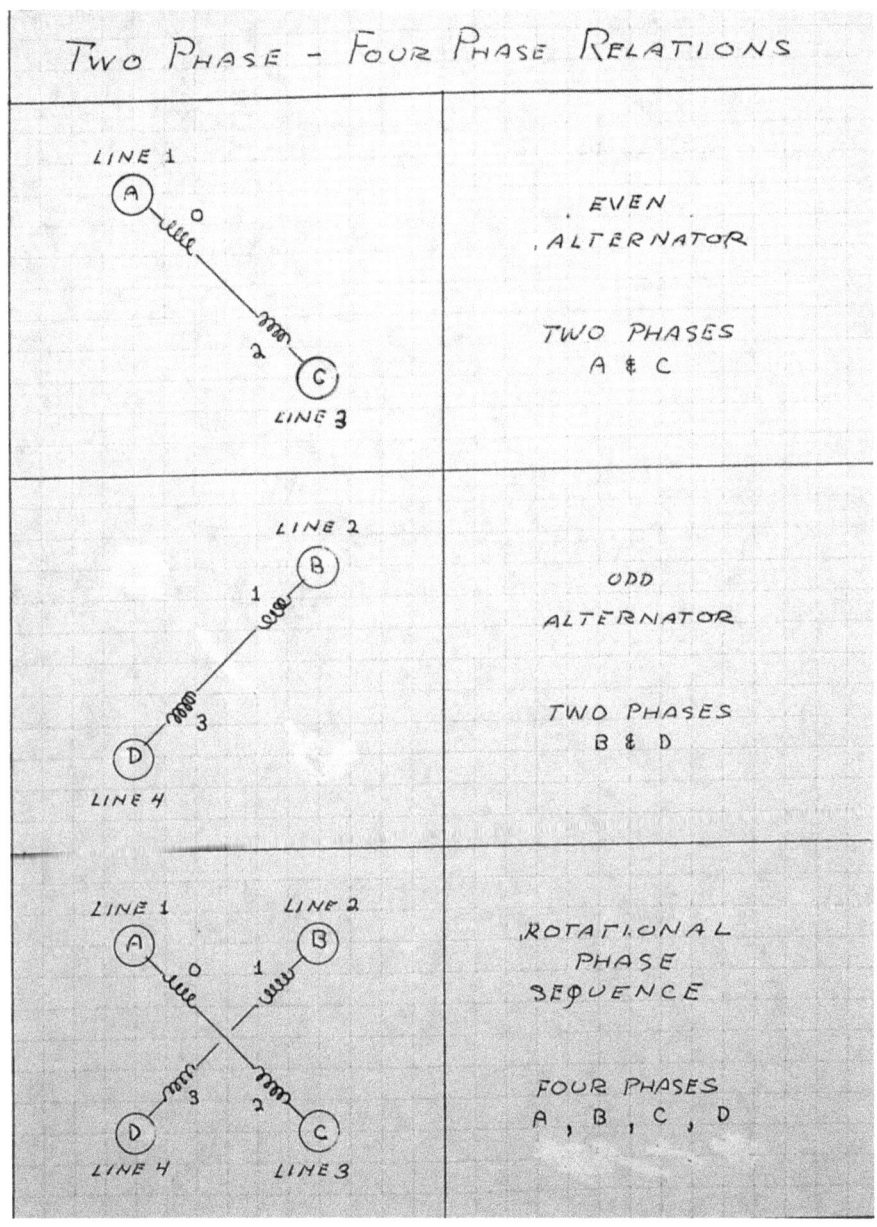

Fig. 16 – Relations of Two Phase and Four Phase

Fig. 17 – Westinghouse alternator in Machinery Building at the 1893 World's Fair

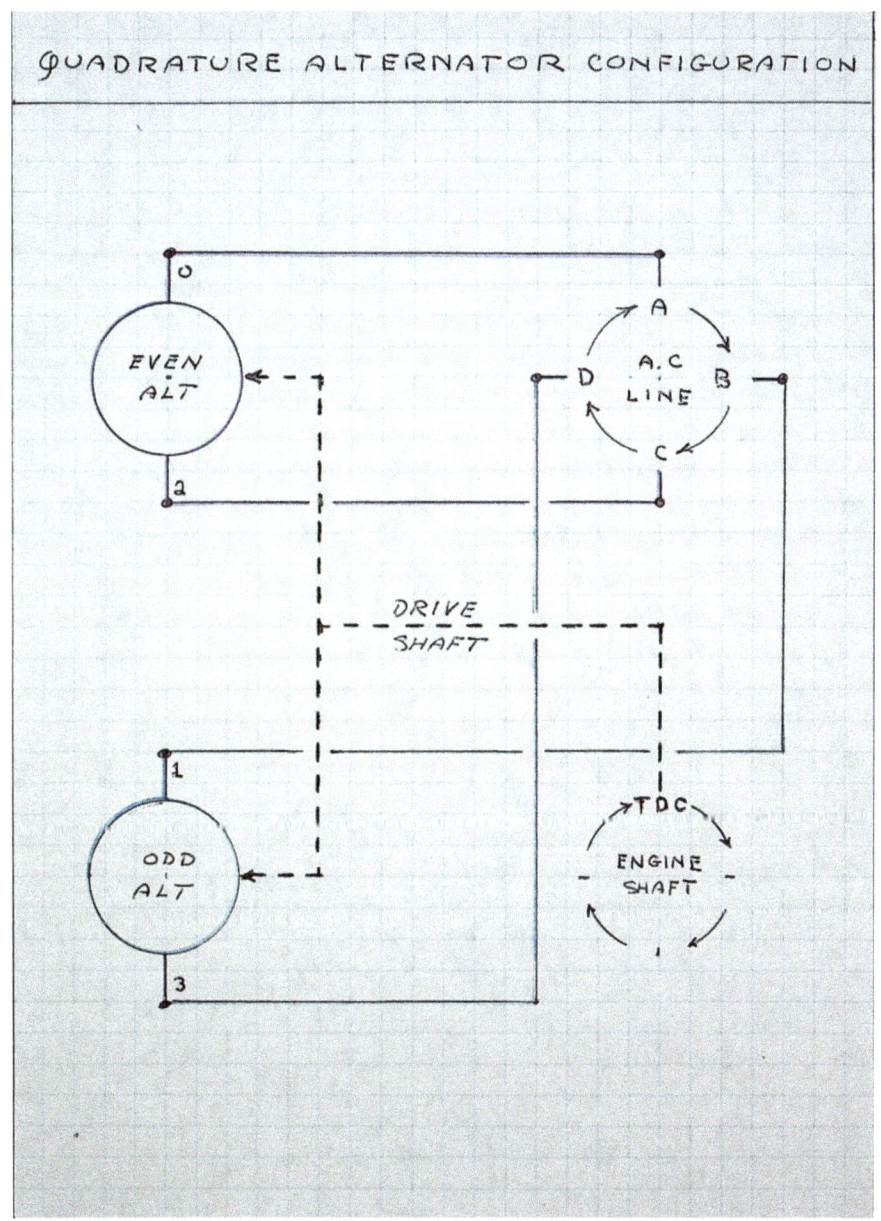

Fig. 17a – Quadrature alternator arrangement

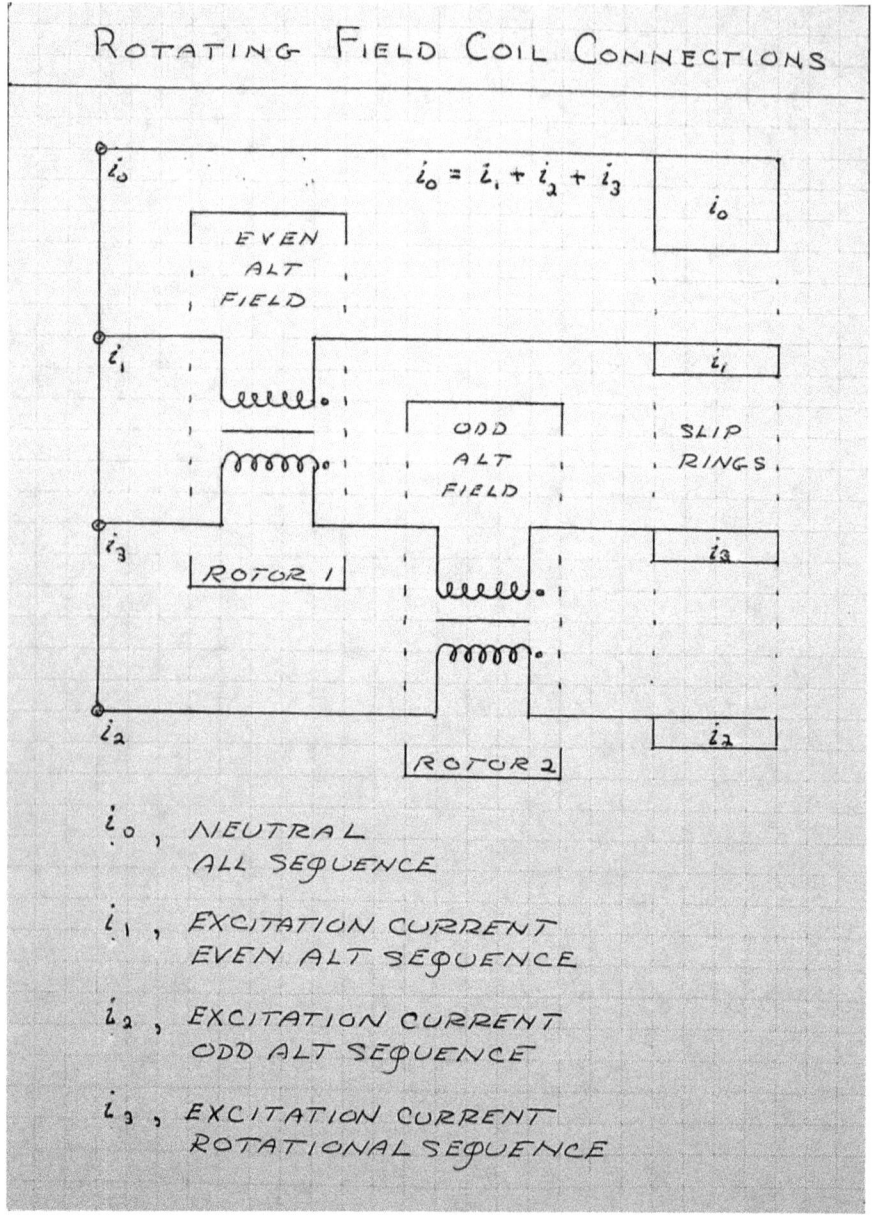

Fig. 18 – Configuration of rotating field coil

Fig. 18a – Westinghouse Polyphase generator

Fig. 18b – Single 750-Kilowatt alternator, showing slip rings and field connections fore and aft

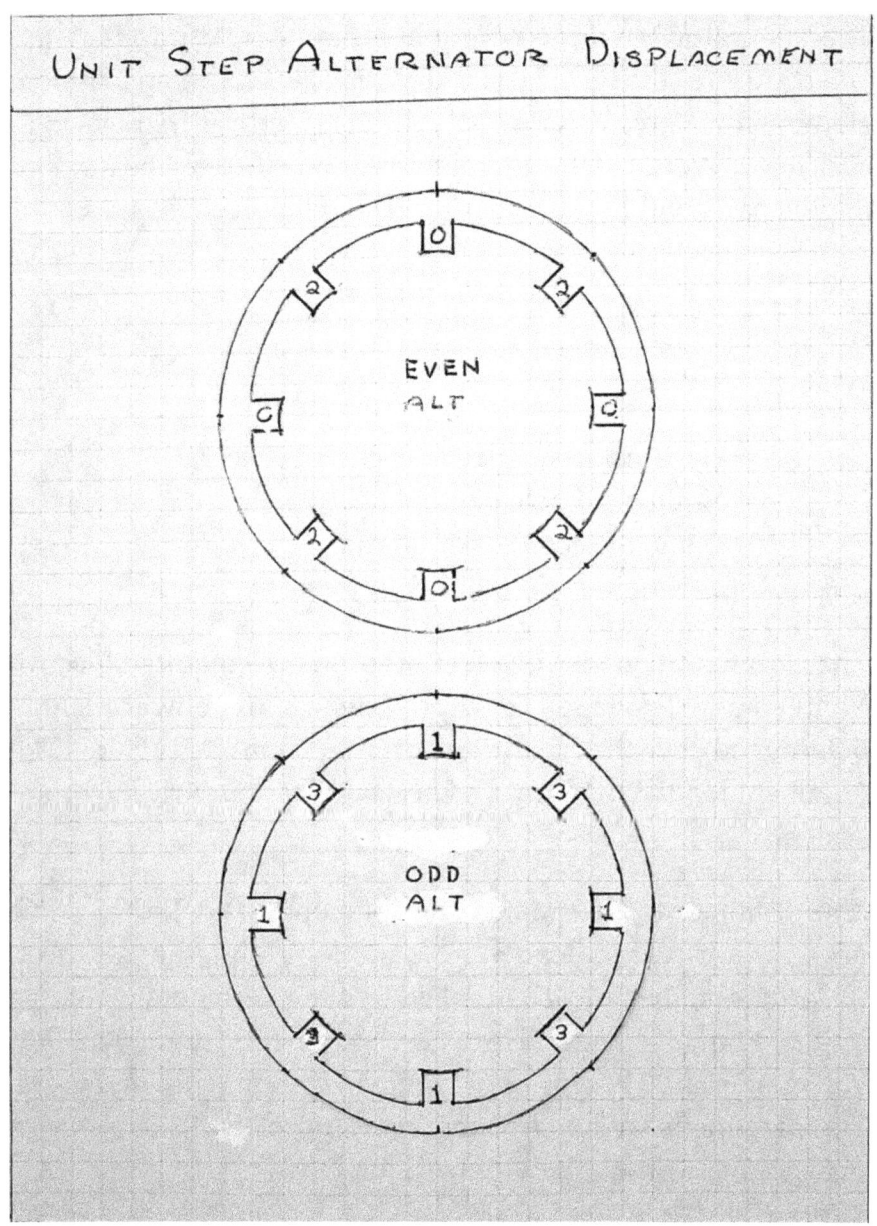

Fig. 14 – Alternator displacement unit step

The alternator armature pole pairs, figure (14) and (14a), are analogous to the gear teeth of a mechanical transmission. Hereby, for a given engine R.P.M., the electrical frequency, in cycles per second, of the alternator is established by the number of pole pairs within the armature. The relation is given as:

$$F = \frac{n}{60} \times RPM$$

Where,

> F = Electrical frequency in cycles per second
> n = Number of pole pairs
> 60 = Ratio of seconds per minute
> RPM = Engine speed in revolutions per minute

The required frequency of alternation, and the required Electro-Motive Force (Voltage), to be provided by the power plant is determined by the character of the connected load. The dominant load at the World's Fair was a vast electrical lighting system.

The distances to be spanned by the connection between the power plant and its load were quite long, thousands of feet. Accordingly, the voltage supplied by the alternators would be in the kilovolt range. The exact voltage is unknown, but voltages of 550, 1100, and 2200, were in common use in this era. The practitioner adage became; one Volt of E.M.F., to span a distance of one foot, the Volt per Foot Law.

Electrical illumination devices, when operating on alternating current, are subjected to an objectionable flicker, that is a pulsation in illumination intensity.

This pulsation presents itself at twice the frequency of alternation, and is most annoying at lower frequencies, 50 cycle per second and lower. Obviously, no flicker exists when operating on direct current.

At this point in history, the most common frequency employed in electrical illumination was 133⅓ cycle per second, this based upon 800 R.P.M. multiples of rotation. Nikola Tesla, however, settled upon 60 cycle per second, this based upon 360 R.P.M. multiples of rotation. This lower, or slower, frequency accordingly favors the operation of electric motors as well as allowing for the design of more economical alternators. Either 133⅓, or 60, cycles per second may have been employed at the World's Fair.

A single electric lamp, having only a single pair of terminals is inherently an alternate sequence device, hence the flicker effect. Individually, these lamps do not require Polyphase power for their operation. However, when groups of lamps are connected in a Polyphase manner, the net result is the cancellation of the flicker effect. This is because during the pulsation of the intensity, a lamp operating on phase, A, may be in the dimming part of its pulsation, while a second lamp operation on phase, B, will in turn be operating in the brightening part of its pulsation. Accordingly, the illumination intensity of this pair of lamps will remain relatively constant, thus the cancellation of the flicker effect. This allows for a somewhat lower frequency of operation, which is particularly effective in the operation of vapor (arc) lamps since these devices are very prone to the flicker effect.

It is estimated that approximately 8,000 kilowatts (11,000 H.P.) of electrical power, this generated by the Westinghouse-Tesla

Alternators, was required to provide the illumination of the 1893 World's Fair. The Power Plant Switchboard, shown in figures (19, 20, 21, 22), regulates and directs this electric power produced by the various alternators.

It is evident from figures, (19) through (22), shown that this switchboard is divided into two decks, a lower forward, deck, and an upper, after deck. The upper deck consists of a row of forty identical panels, while the lower decks consist of five simple panels on the left side, and thirteen complex panels to the right. Four field rheostat dials present themselves on four of the larger panels, and thirteen field rheostat dials present themselves on each of the smaller panels. These rheostats serve to regulate the power produced by their corresponding alternators. Accordingly, a total of seventeen alternators are controlled by this switchboard.

Examination of the photographs which portray the switchboard would indicate the lower deck group of thirteen panels directly serve the same number (13) of Two-Phase feeders.

Fig. 19 - Westinghouse Alternating Current switch board

Fig. 20 - Westinghouse Alternating Current switch board

Fig. 21 - Westinghouse Alternating Current switch board

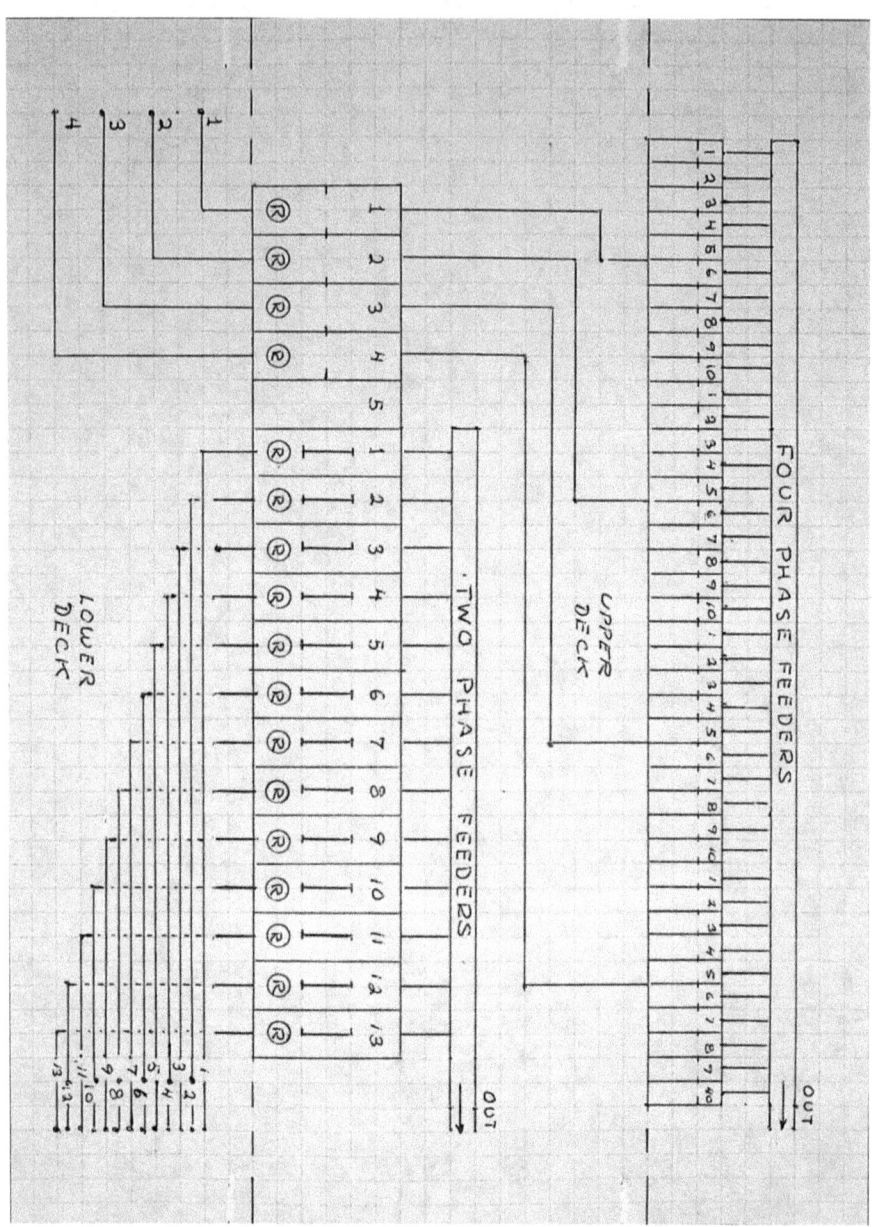

Fig. 22 – Westinghouse switchboard arrangement

These feeders in turn extend to the outside planet. However, by inspection, it is seen that the lower deck group of four panels have no related feeder apparatus. Thus, it may be inferred that the related feeder apparatus is the upper deck group of forty panels. These serve the same number (40) of Four Phase feeders which in turn extend to the outside plant. It can also be inferred that the lower deck group of four panels control four large direct engine driven alternator sets, figure (23).

The primary function of the Power Plant is to supply the required electric power to illuminate the 1893 Chicago World's Fair. Two methods of electrical illumination had come into common use at this point in history;

1. The conductive filament incandescent lamp, these connected in multiple (parallel), and operating from a low voltage, constant potential source of electric power.
2. The carbon vapor arc lamp, these connected in series, and operating from a high voltage, constant current source of electric power.

In all likelihood, the 40 upper deck panels and their related alternators served the parallel connected incandescent lamps, and were regulated for constant potential, the current in proportional variation with the magnitude of the applied load. Likewise, the thirteen lower deck panels and their related alternators served the series connected arc lamps and were regulated for a constant current (Amperage), the potential (Voltage) in proportional variation with the magnitude of the applied load.

Fig. 23 – Alternators in "Machinery Hall"

The installation of nearly one hundred thousand electric lamps served to illuminate the entirety of the 1893 World's Fair. Accordingly, a vast array of electric power feeders extended outward from the switchboard terminals to the various points of distribution and utilization. These feeders were carried throughout by an extensive arrangement of underground passageways, figure (24).

Two distinct categories of electric power distribution present themselves at this installation;

1. That of constant potential (Volts) serving a complex arrangement of load centers with their related "step down" transformers, which in turn deliver electric power to the parallel connected incandescent lamps, figure (25).
2. That of constant current (Amperes) serving simple arrangement of series connected arc lamps, figure (26).

In the parallel connected constant potential system the current in Amperes varies in direct proportion to the number of incandescent lamps involved, figure (27), whereas, in the series connected constant current system the potential in Volts varies in direct proportion to the number of arc lamps involved, figure (28).

The incandescent lamp illumination operates at a high current with a correspondingly low voltage, employing the "Stanley Transformer" to transform conditions from the high voltage, low current delivered by the switchboard. Hereby, a system relatively safe for human contact is presented.

Fig. 24 – Method of underground electric transmission at the 1893 World's Fair

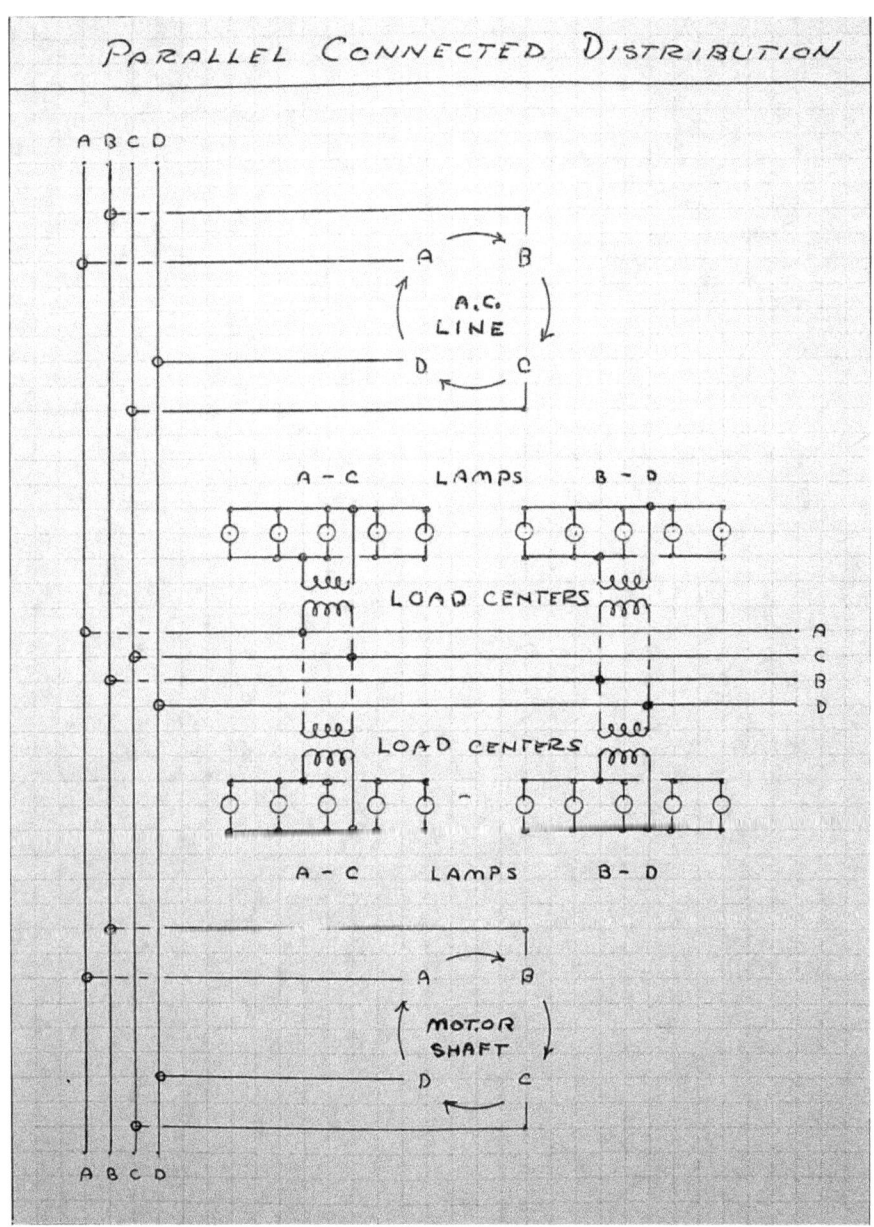

Fig. 25 – Method of parallel distribution

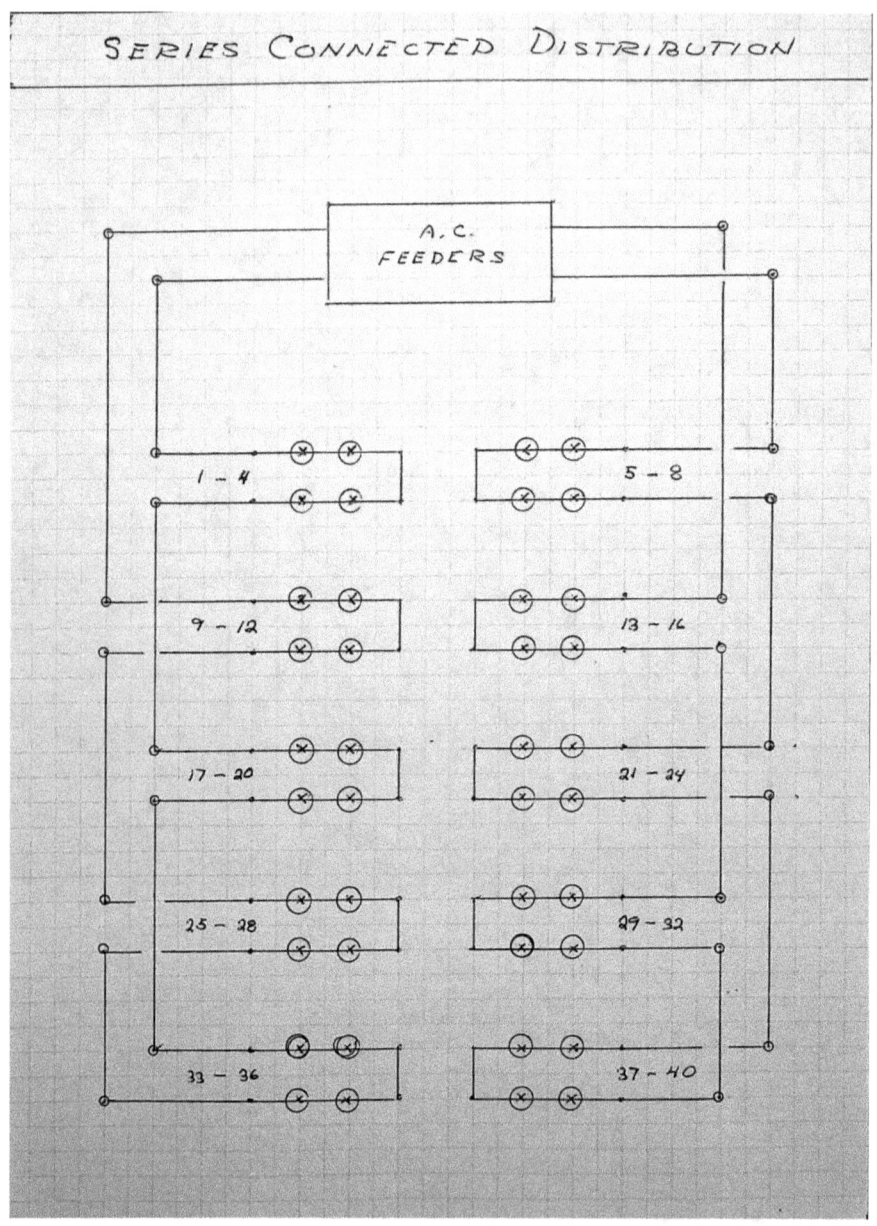

Fig. 26 – Method of series distribution

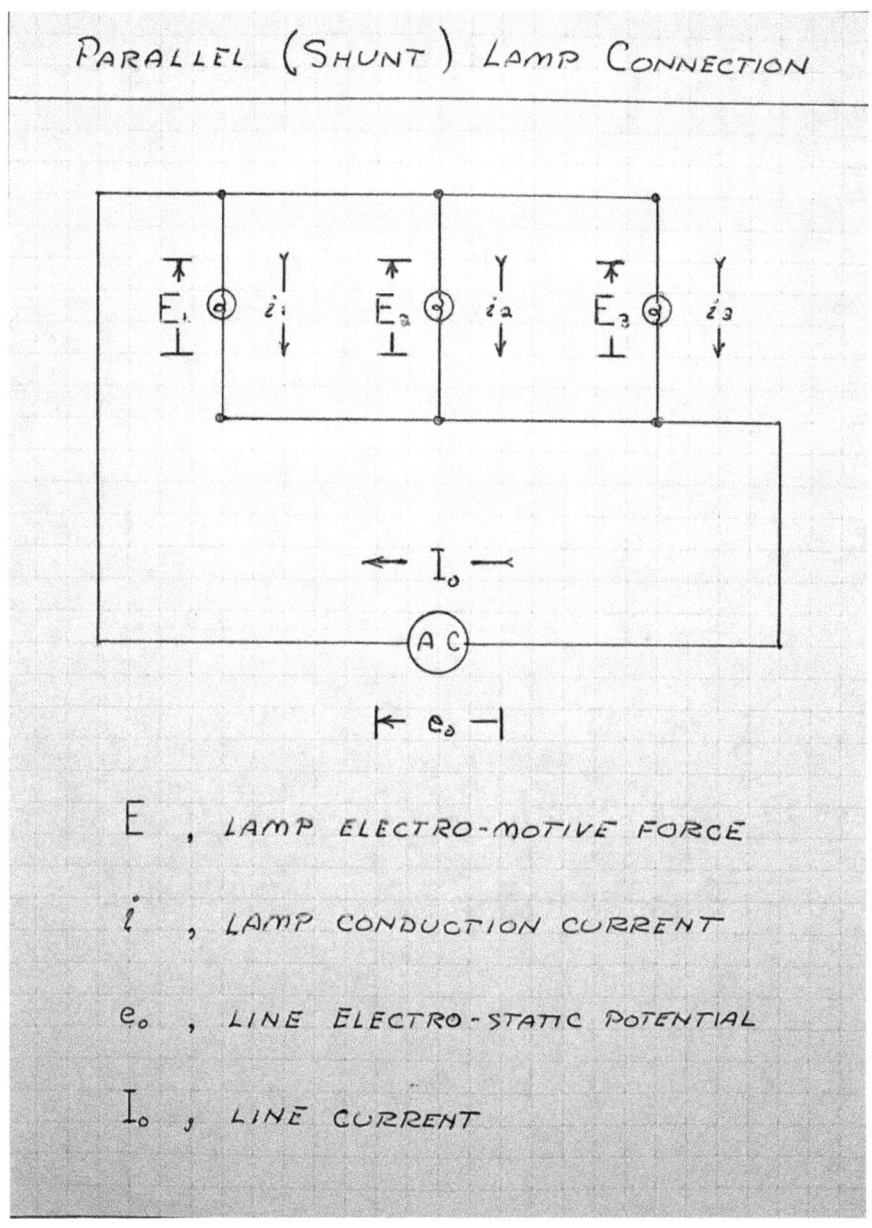

Fig. 27 – Parallel electric lamp connection

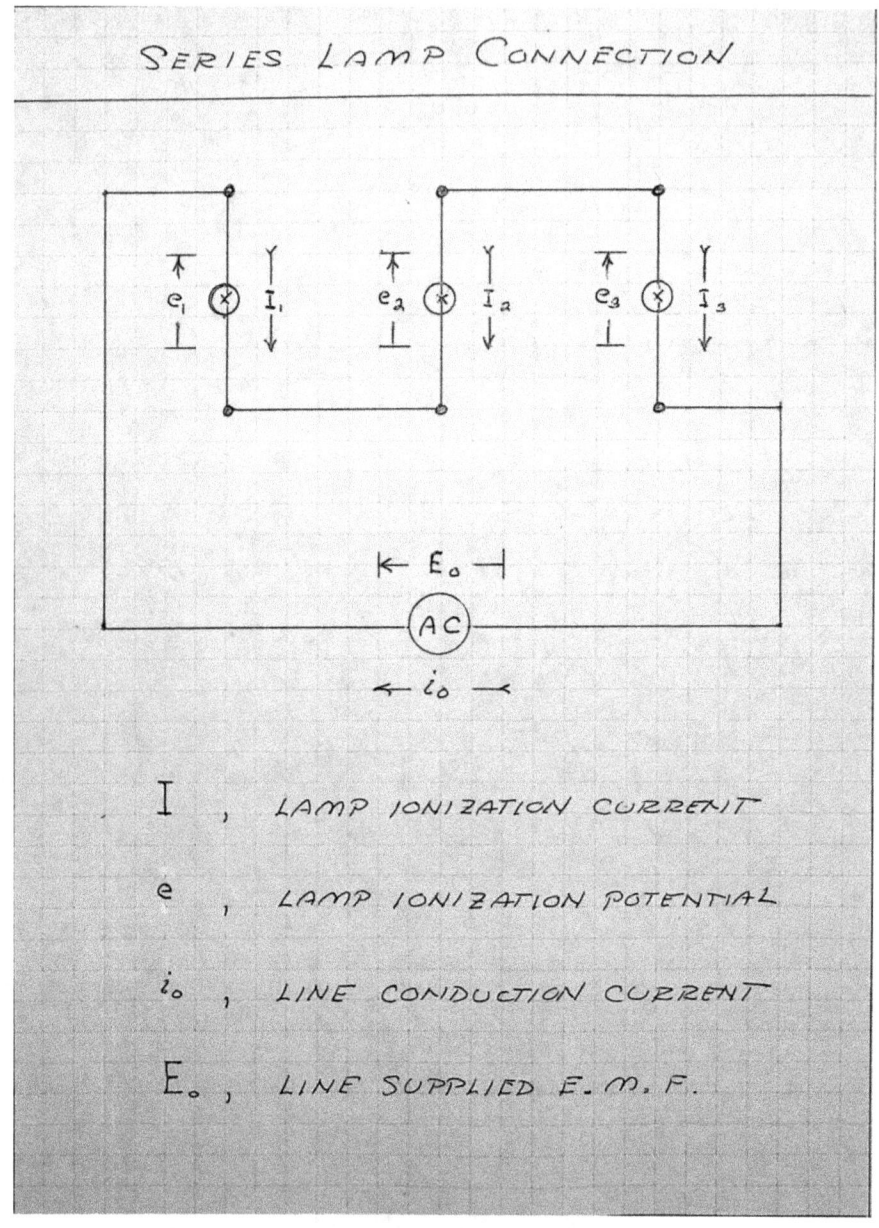

Fig. 28 – Series electric lamp connection

Conversely, the arc lamp illumination operates at a high voltage with a correspondingly low current, this without the employment of transformers. Accordingly, this system is dangerous for human contact. Consequently, incandescent lamp illumination is well suited for indoor operation in close proximity to human activity, whereas the arc lamp illumination is predominantly an outdoor system, or indoors far from human reach, in either case operating remotely from human activity.

A practical example of the comparison of parallel versus series operation is presented by figure (29). In this example three identical incandescent lamps are employed. Each lamp is rated at an Electro-Motive Force of 110 Volts, producing a conduction current of 0.11 Ampere, or conversely, rated at 0.11 Ampere, producing an Electrostatic Potential of 110 Volts.

It is evident from figure (29) and (29a), that for the parallel connection of these lamps the applied voltage is identical across all three lamps, 110 Volt, but the current drawn by them together is additive, 0.33A or 330 milliampere. The total electric power consumed by this configuration of lamps is the product of the applied voltage, 110V, and the total current, 330mA, that is, about 36 Volt-Ampere, or Watt.

Likewise, the apparent electric conductance offered by this connection of lamps to the source of electric power is the ratio of the total current drawn, 330mA, to the applied voltage, 110V, that is, 1.0 millisiemens.

12 WATT , 110 VOLT , 0.11 AMPERE LAMPS

PARALLEL LAMP CONNECTION

$E = 110$ VOLT $\qquad i = 0.11$ AMPERE

$G = i/E = 0.001$ SIEMENS

$P = i \cdot E = 12.1$ WATT

$e_o = 110$ VOLT $\qquad I_o = 0.33$ AMPERE

$P_o = e_o I_o = 36.3$ WATT

SERIES LAMP CONNECTION

$I = 0.11$ AMPERE $\qquad e = 110$ VOLT

$R = e/I = 1000$ OHM

$P = e \cdot I = 12.1$ WATT

$i_o = 0.11$ AMPERE $\qquad E_o = 330$ VOLT

$P_o = i_o E_o = 36.3$ WATT

Fig. 29 & 30 – Electrical characteristics of electric lamp configurations (Figure 30 is presented in the lower half of the image)

CONSTANT POTENTIAL RELATIONS

$$e_0 = \text{CONSTANT MAGNITUDE}$$

(1) $\quad e_0 = E_1 = E_2 = E_3 \qquad \text{VOLT}$

(2) $\quad I_0 = i_1 + i_2 + i_3 \qquad \text{AMPERE}$

(3) $\quad P_0 = e_0 I_0 \qquad \text{VOLT-AMPERE}$

(4) $\quad P_0 = P_1 + P_2 + P_3 \qquad \text{WATT}$

$$P_1 = i_1 E_1 \qquad P_2 = i_2 E_2 \qquad P_3 = i_3 E_3$$

Fig. 29a – Relations of constant potential

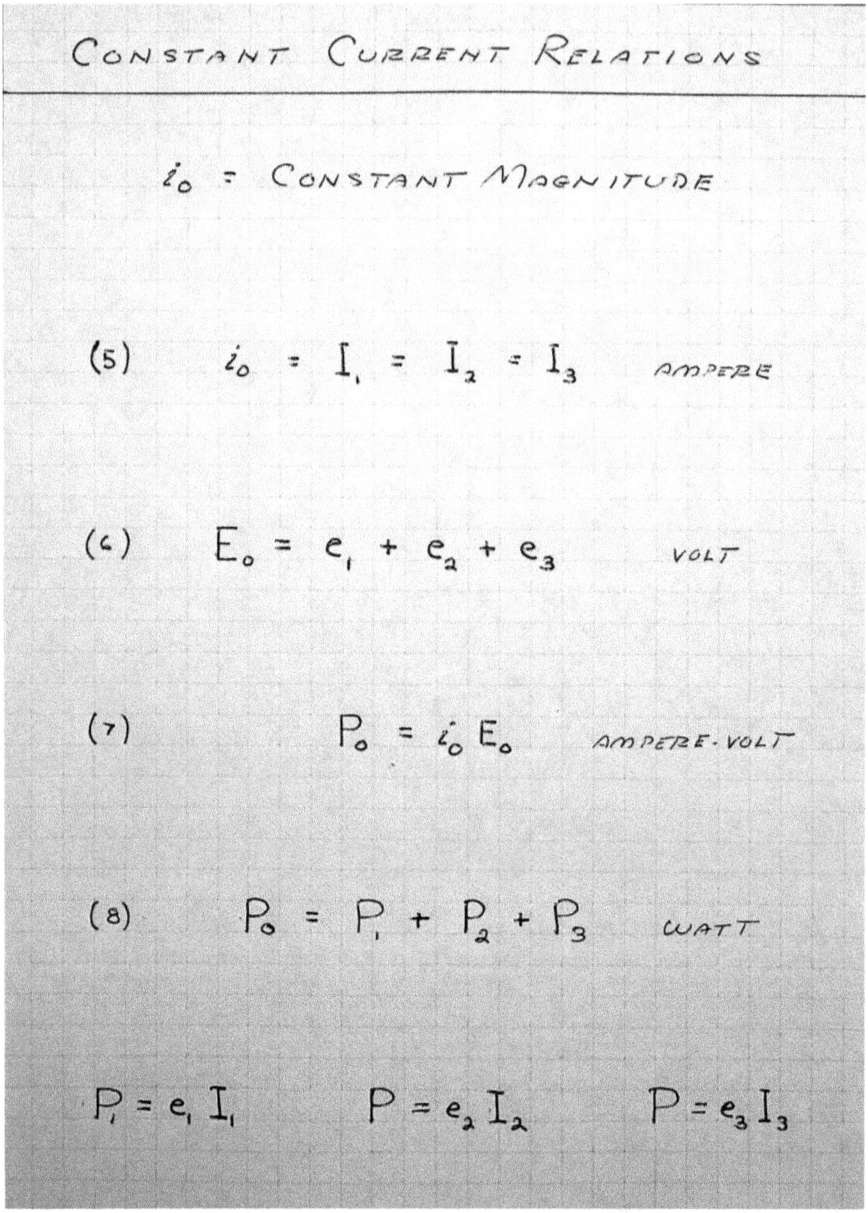

CONSTANT CURRENT RELATIONS

$$i_0 = \text{CONSTANT MAGNITUDE}$$

$$(5) \qquad i_0 = I_1 = I_2 = I_3 \qquad \text{AMPERE}$$

$$(6) \qquad E_0 = e_1 + e_2 + e_3 \qquad \text{VOLT}$$

$$(7) \qquad P_0 = i_0 E_0 \qquad \text{AMPERE·VOLT}$$

$$(8) \qquad P_0 = P_1 + P_2 + P_3 \qquad \text{WATT}$$

$$P_1 = e_1 I_1 \qquad P = e_2 I_2 \qquad P = e_3 I_3$$

Fig. 30a – Relations of constant current

Conversely, it is also evident from figure (30) and (30a), that for the series connection of these lamps the applied current is identical through all three lamps, 0.11 Ampere, or 110 milliampere, but the Electro-Motive Force produced by them is additive, 330 Volt. Again, the total electric power consumed by this configuration of lamps is the product of the applied current, 110mA, and the total Electro-Motive Force, 330V, that is, about 36 Ampere-Volt, or Watt.

Likewise, the apparent electric resistance offered by this connection of lamps to the source of electric power is the ratio of the total applied Electro-Motive Force, 330V, to the total current drawn, 110mA, that is, 1.0 Kilo-Ohm.

Hence, the electric power consumed by either the parallel, or series, connection of these three lamps is identical, 36 Watts, since the Volt-Ampere, or Ampere-Volt products are equivalent. However, their ratios, Ampere per Volt, or Volt per Ampere, presented to the source of electric power are reciprocal with respect to each other.

Hereby, through a basic analysis of the somewhat primitive arrangements utilized at the 1893 World's Fair, the most fundamental laws of electricity have been derived.

References

[1]   Steinmetz: Engineer and Socialist, Ronald R. Kline, 1992, Page 268

[2]   Electromagnetic Theory, Oliver Heaviside, 1950, Page xxii
Steinmetz: Engineer and Socialist, Ronald R. Kline, 1992, Chapter 1, Pages 6-11

[3]   Theory and Calculation of Alternating Current Phenomena, C. P. Steinmetz, 1900, Chapter 5, Pages 33-36

[4]   Steinmetz: Engineer and Socialist, Ronald R. Kline, 1992, Chapter 4, Pages 89-91, Chapter 7, Page 189

[5]   Steinmetz: Engineer and Socialist, Ronald R. Kline, 1992, Chapter 4, Page 86, Chapter 6, Page 115, Pages 137-138

[6]   Steinmetz: Engineer and Socialist, Ronald R. Kline, 1992, Chapter 26, All Pages
Steinmetz: Engineer and Socialist, Ronald R. Kline, 1992, Appendix, Article 305, Page 495

[7]   General Electric Review Vol. 39 No.12, E.C. Sanders, 1936, Pages 962-964

[8]   Method of Symmetrical Co-Ordinates Applied to The Solution of Polyphase Networks, Charles L. Fortescue, 1918, Page 1027

[9]   Symmetrical Components, C.F. Wagner & R.D. Evans, 1926, Chapter 1, Article 3, Page 6

[10]   Symmetrical Components, C.F. Wagner & R.D. Evans, 1926, Introduction, Page 15

[11]   Method of Symmetrical Co-Ordinates Applied to The Solution of Polyphase Networks, Charles L. Fortescue, 1918, Pages 1116-1129

[12]   Symmetrical Components, C.F. Wagner & R.D. Evans, 1926, Chapter 1, Article 2, Page 5

[13]   Theory And Calculation of Alternating Current Phenomena, C. P. Steinmetz, 1900, Chapter 5, Article 27, Pages 36-38

Appendix

Alternators

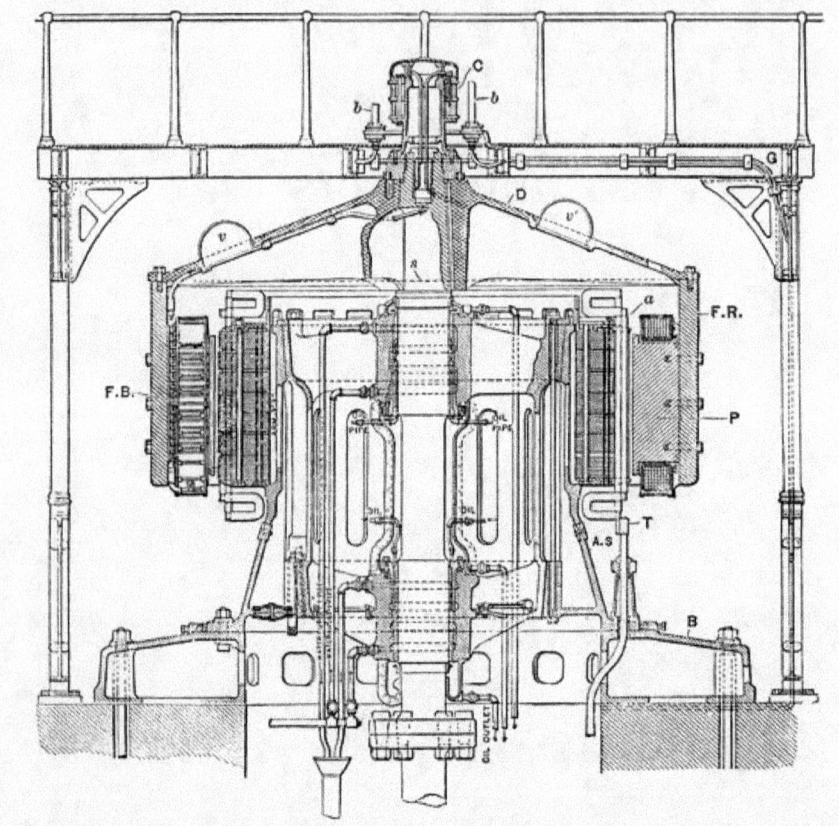

FIRST 5000 HORSE-POWER NIAGARA ALTERNATOR AS DESIGNED AND INSTALLED

Fig. a – Niagara Falls Generating Plant alternator diagram

Fig. b – Alternator assembly

Fig. c – Installed alternator at Niagara

Turbines

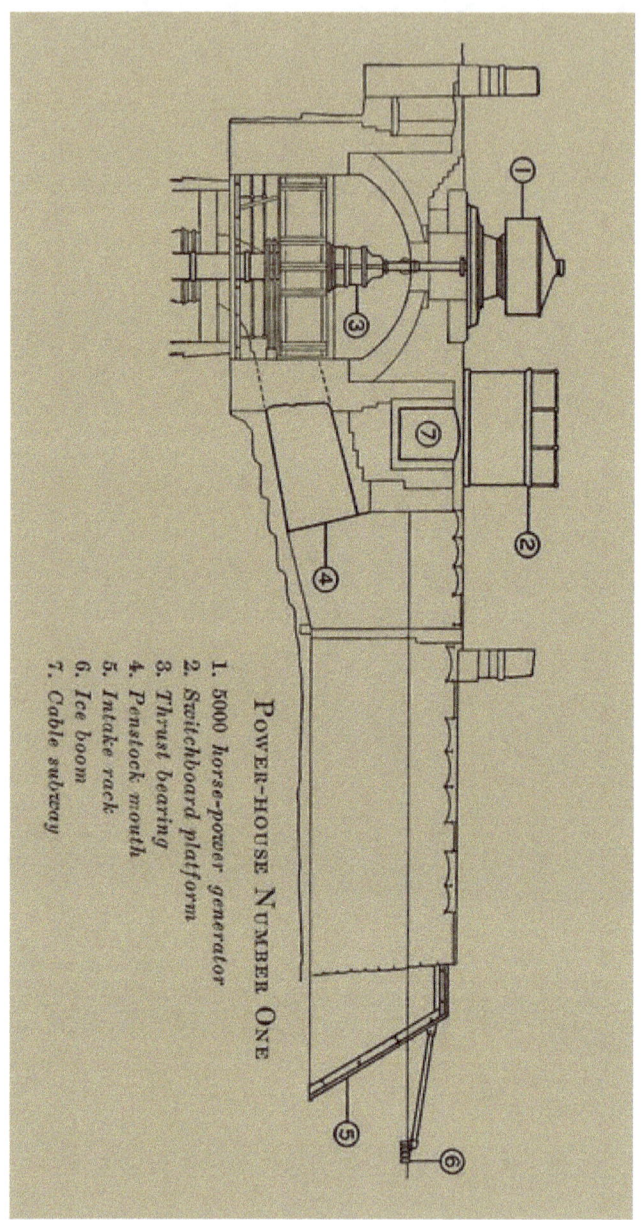

Fig. d – Side-view of Niagara Station No. 1 layout

GENERATING STATION OF THE NIAGARA FALLS POWER COMPANY, SHOWING THE TEN 5,000 H. P. GENERATORS

Fig. e – 5,000 horsepower generators

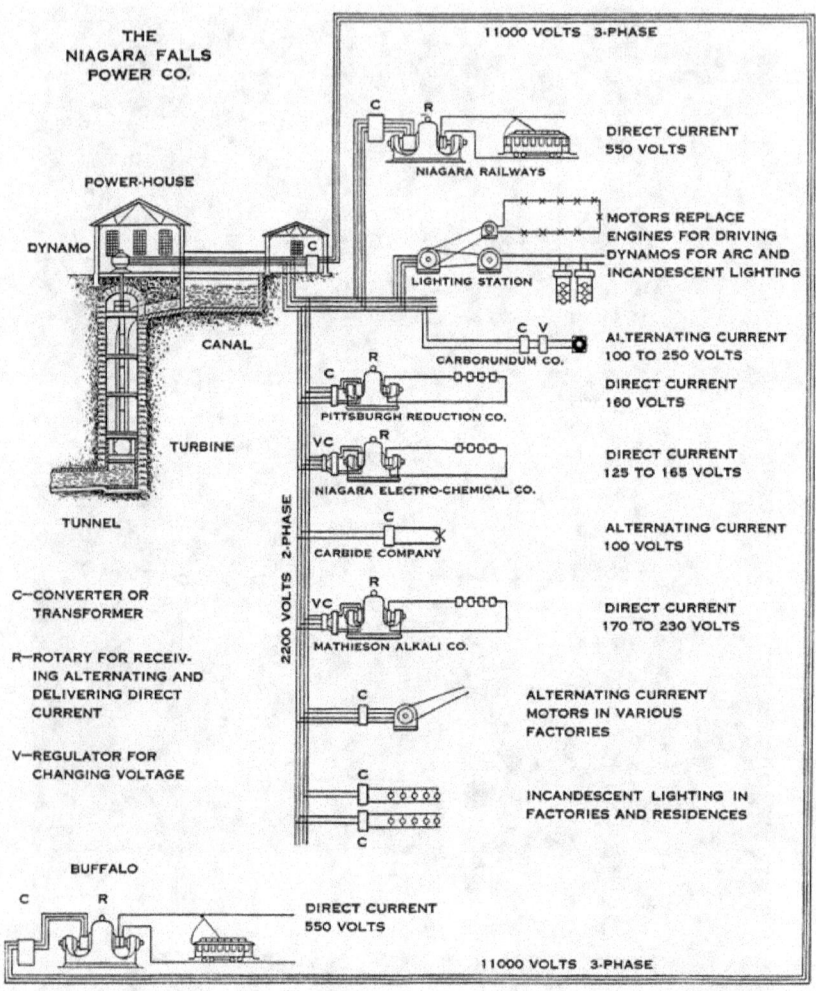

Fig. f – Niagara Falls Generating Plant layout

Transmission & Distribution

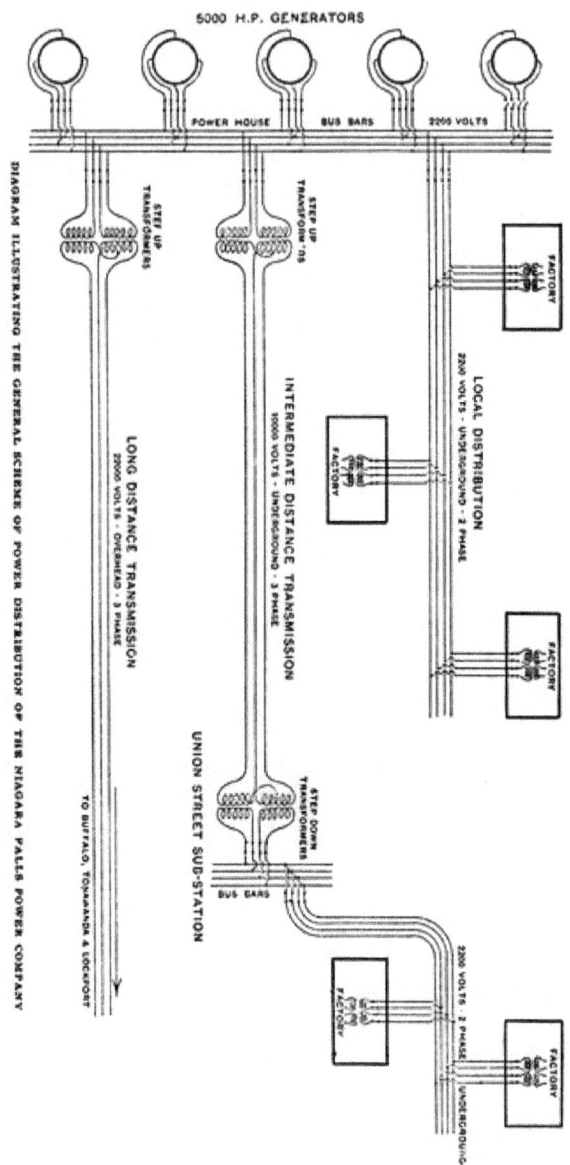

Fig. g – Power distribution scheme at the Niagara Power Plant

AIR-BLAST TRANSFORMERS AT NIAGARA FOR THE EARLY TRANSMISSION TO BUFFALO

Fig. h – Original air-blast transformers for power transmission

Fig. i – Closeup of a dual air-blast transformer unit

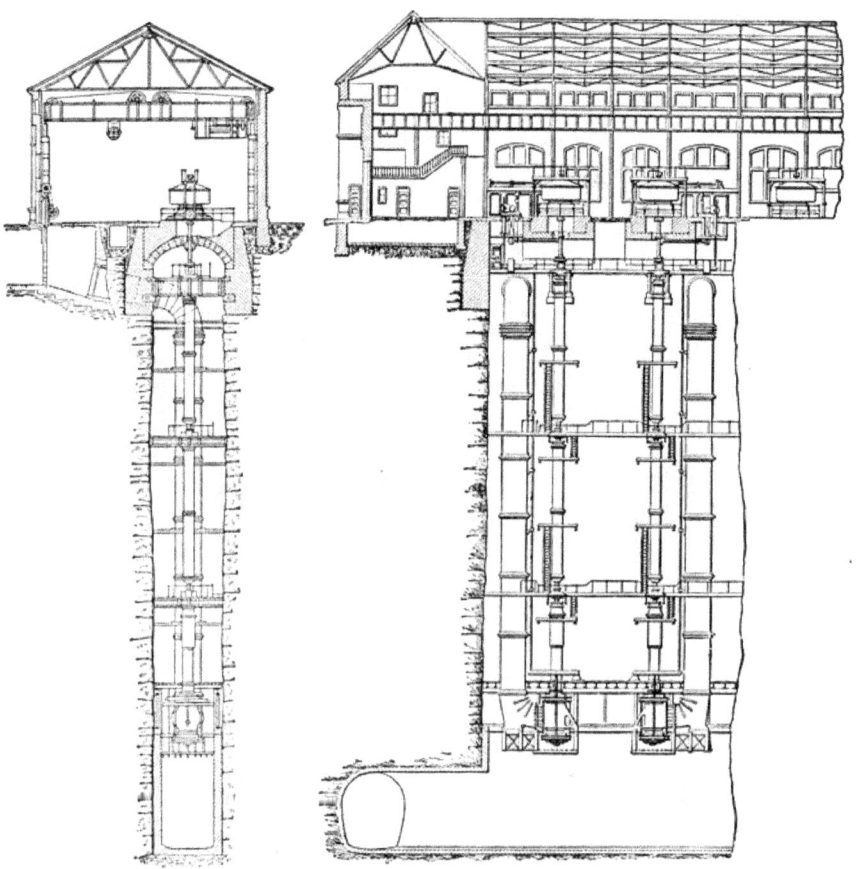

Fig. j – Turbine and alternator configuration at Niagara Falls

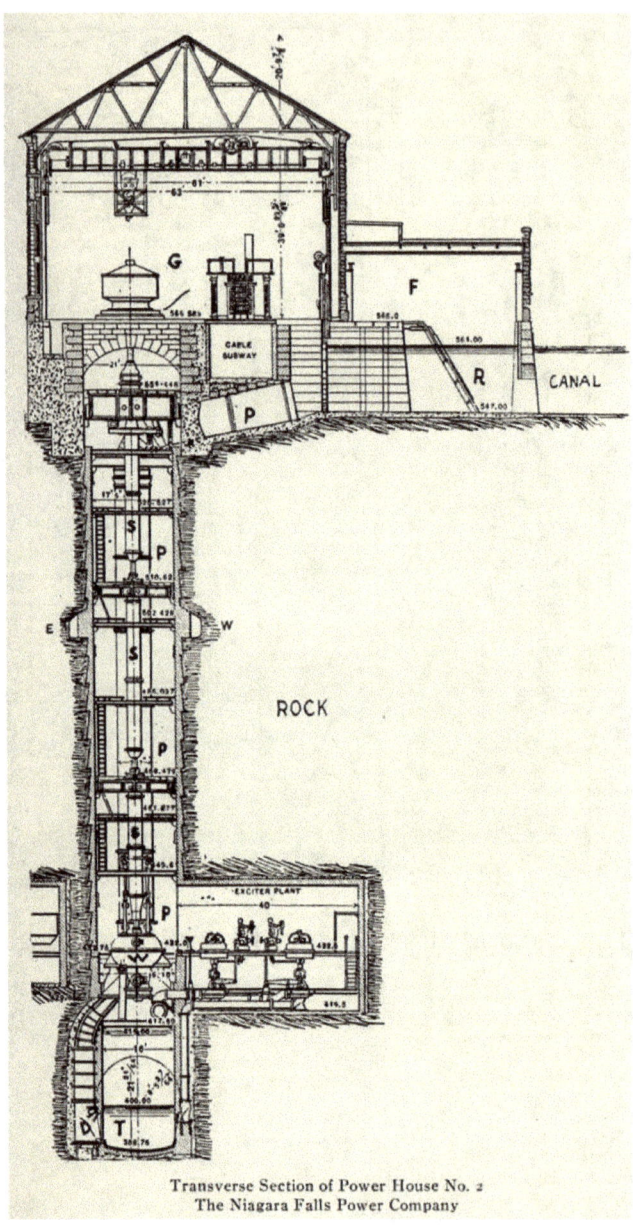

Transverse Section of Power House No. 2
The Niagara Falls Power Company

Fig. k – Niagara Falls Generating Powerhouse No. 2

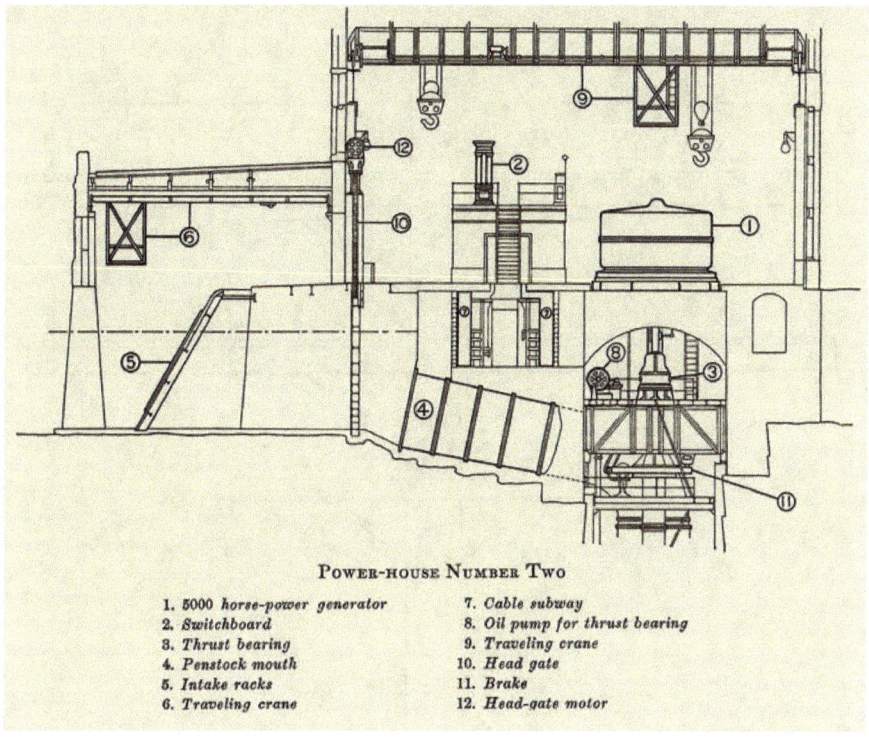

POWER-HOUSE NUMBER TWO

1. 5000 horse-power generator
2. Switchboard
3. Thrust bearing
4. Penstock mouth
5. Intake racks
6. Traveling crane

7. Cable subway
8. Oil pump for thrust bearing
9. Traveling crane
10. Head gate
11. Brake
12. Head-gate motor

Fig. 1 – Cross-sectional view of Powerhouse No. 2

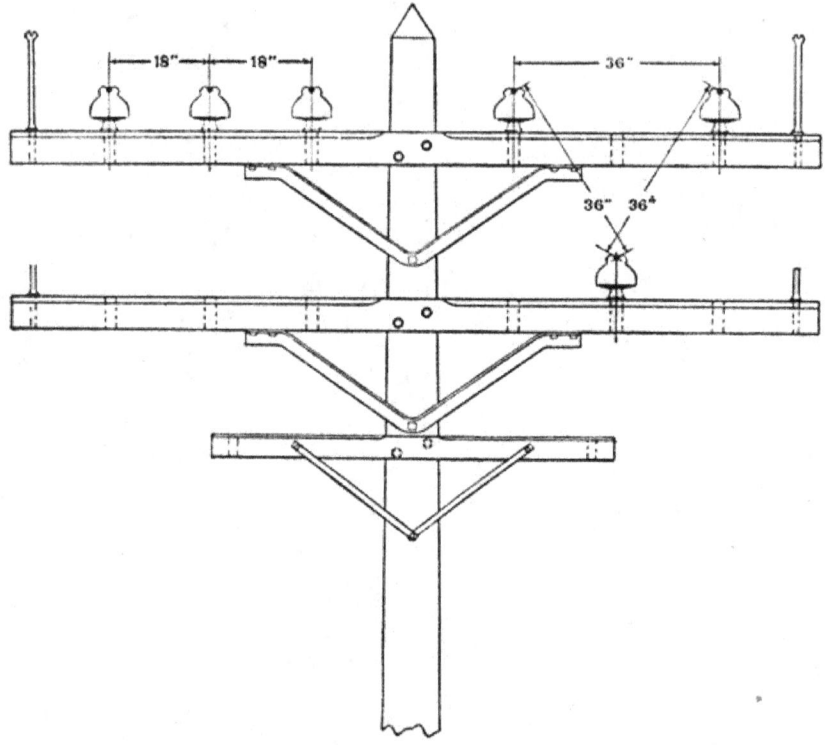

Fig. m – Crossarm arrangement for Niagara to Buffalo electric power transmission

Fig. n – Construction of alternator shaft

Fig. 0 – Adams Power Plant Transformer House

Fig. p – Turbine Hall of the Niagara Power House

Fig. q – Turbine Hall of the Niagara Power House with switchboard in view

Fig. r – Turbine Hall of the Niagara Power House

Fig. s – The Niagara Falls Power Company electric transmission right-of-way towards Buffalo

Historical Texts

'Fast tracking' a building where the foundation is completed before the design of the upper floors is complete is a common current practice. This principle was applied to construction of the Adams Station. Because a tailrace tunnel was common to any project, a contract was awarded in September 1890. Construction of the 1-1/4 mile long tunnel commenced in October 1890 and was completed in December 1892 [Fig. 4.1]. A horseshoe shape 21 feet high and 18.8 feet wide was selected [Fig. 4.2]. During construction through the weak Rochester shale strata, it became necessary to line the tunnel using a total of sixteen million bricks in four courses set in Portland cement. A profile of the tunnel shows the extent of the shale [Fig. 4.3].

By July 1891 plans had reached the following stage:[41]

1. Hydraulic system consisting of a single inlet canal, the shortest possible tailrace tunnel, and a central station for power development. The turbine units should be large and mounted at the bottom of the wheel pit. Their complementary machines, whether pumps, compressors or generators should be mounted on the top of the same rotating shaft so as to constitute a unit of power. [Fig. 4.4]

2. Local distribution of power by direct current of electricity.

3. Transmission of power to Buffalo by compressed air.

The inlet canal was started in August 1891 and completed in October 1892 [Figure 4.5]. The 20 feet wide and 182 feet deep stone masonry wheel pit was started in late 1891 and completed in January 1894 [Fig. 4.6].[42]

The state of the art of electric power transmission in the early 1890's was as follows:[43]

1. 1890 - Willamette Falls to Portland, Oregon - 4,000 volts ac single-phase 12 miles for lighting.

2. 1891 - Telluride, Colorado - 3,000 volts ac single-phase three miles supplying a 100-hp synchronous motor for operating an ore crushing plant.

3. 1891 - Lauffen to Frankfurt, Germany - Experimental 30,000 volt three-phase line 108 miles for 300 hp for the Frankfort Exposition. Overall efficiency 77 percent.

In early autumn of 1891 it became evident that alternating current could be safely and economically controlled for the transmission of power more than five times the distance from Niagara to Buffalo. From this period, all serious attention was concentrated upon electrical installations; generators, transformers, transmission lines, motors, and power and light distribution.[44]

Fig. t

In December 1891 an invitation to submit proposals for generation of electrical energy for local lighting and power purposes was sent to three United States and three Swiss electrical equipment designers and manufacturers. The invitation did not mention direct current, alternating current or voltage. It took a year until all proposals were received.[45]

In 1892 Professor George Forbes, an electrical engineer from England, was hired as a consultant [Fig. 4.7]. He had submitted the polyphase alternating current electrical proposal to the International Niagara Commission contest.[46] Horsepower and speed were determined by consultation with the hydraulic turbine and electrical equipment manufacturers.[47] The contemplation of the use of electricity from hydraulic units of 5000 hp in a plant, which would ultimately aggregate 100,000 hp, required imagination, optimism and confidence. 100,000 hp approximated the output of all the electric lighting stations in the United States.[48]

In July 1892 a contract was awarded to Faesch & Piccard of Geneva, Switzerland for complete working drawings of a 5000-hp 250-revolutions-per-minute hydraulic turbine including the governor.[49] The turbines far exceeded in power and speed any then in existence.[50] Transportation costs and import duties made domestic manufacture economical.[51] In November 1892 a contract was awarded to I. P. Morris of Philadelphia for two turbines. A third was added in 1893.[52] The turbines were double runner Fourneyron type with outward discharge directly into the wheel pit [Fig. 4.8 & 4.9]. There were no draft tubes. Water pressure on the top runner supported a large portion of the weight of the revolving parts of the units, the remainder being carried by a collar thrust bearing.[53] The mechanical governors, which were made in Switzerland by Faesch & Piccard, regulated the flow of water by raising or lowering a circular collar placed outside the turbine discharge.[54] The variation in speed would not exceed two percent normally or four percent with a 25 percent variation in load.[55] This speed variation would be totally unacceptable today.

The New York City architectural firm of McKim, Mead & White was engaged to design the limestone powerhouse [Fig. 4.10].[56]

Fig. u

It is interesting that in late 1892 the turbines were ordered four months before all the proposals were received for generators for local lighting and power. This was the era of the 'Battle of the Currents' with Edison as the leading proponent of direct current and with Westinghouse and others promoting alternating current. When the first execution by electricity took place in 1890 in New York State's Auburn Prison, an ac electric chair was used. Dc proponents said the condemned murderer had been "Westinghoused."[57]

The opportunity for a large scale demonstration of the alternating current system to dispel fears about the system's high voltages and display its versatility came in the spring of 1892 when bids were taken to light the grounds of the Columbian Exposition to be held in Chicago in 1893 [Fig. 5.1]. Westinghouse astounded everyone by bidding about one-third of the bid submitted by the recently formed General Electric Company. GE held the Edison incandescent lamp patents, a seeming crippling handicap. However, Westinghouse owned the rights to an 1880 Sawyer-Man patent for a two-piece lamp, which had a glass globe and a cork-like ground glass stopper that didn't infringe on the Edison patents [Fig. 5.2]. In less than a year Westinghouse built a factory and produced a quarter million of these "stopper" lamps [Fig. 5.3].

The Exhibition used more electricity than the whole City of Chicago. The central-station plant that supplied the fair consisted of twelve two-phase generators rated 750 kW, 60 Hz, 2000 to 2300 volts, belt driven or direct driven by five reciprocating steam engines [Fig. 5.4]. The Westinghouse exhibit in Machinery Hall was the largest plant of the Tesla polyphase system operating at that time. In addition to the lighting plant, there was an exhibit of a complete polyphase system which showed power could be transmitted great distances, and then be utilized for various purposes. Taking power from the 60-Hz fair circuits was a 500-hp motor that drove a two-phase 500-hp, 30-Hz, 400-volt generator Fig. 5.5]. This powered a 400 volt-to-10,000 volt step-up transformer, a short 'transmission' line, and a step-down transformer which ran induction motors, a synchronous motor plus a rotary converter for changing alternating current to direct current.[58] The converter had a rotating armature with four slip rings for the two-phase ac and a commutator for the dc [Fig 5.6].[59]

The Westinghouse polyphase exhibit was a big attraction when the Columbian Exhibition opened in the spring of 1893. Wooden models of the hydraulic turbines and governors under construction for Niagara Falls were also exhibited.[60] Representatives of the Cataract company had attended demonstrations of polyphase equipment at the Westinghouse factory and on May 6, 1893 adopted polyphase alternating current for both local use and transmission to Buffalo.[61]

Fig. v

Complications of the electrical design delayed awarding a contract for the generators. Transportation costs, import duties and patent questions removed the foreign manufacturers from competition.[62] The generator designs submitted in March 1893 by Westinghouse and General Electric did not fulfill the conditions imposed by the turbine designers for angular momentum and limited weight. On August 10, 1893 new bids were requested based on a design by Professor Forbes of the Cataract company for a 20,000 volt two-phase generator with an external rotating field of the "umbrella" type whereas a rotating armature was standard practice [Fig. 5.8].[63] Figure 5.9 shows the rotating field ring with the field poles and field windings. The armature winding was stationary. Westinghouse proposed two-phase; General Electric recommended three-phase.[64] Two-phase four-wire was selected because it was expected that much of the local load would be single-phase.[65] Next the frequency had to be selected. Hydraulic turbines had been ordered with a speed of 250 rpm. The frequencies possible at 250 rpm were: 16 2/3-Hz with 8 poles on the rotating field; 25-Hz with 12 poles; 33 1/3-Hz with 16 poles and 41 2/3- Hz with 20 poles. Low frequencies were preferred for large motors and rotary converters. Professor Forbes preferred 16 2/3- Hz for the commutating type ac motors then in use. Higher frequencies were more suitable for incandescent and arc lights. Tests had shown that at 25 Hz incandescent lamps did not show objectionable flickering. Westinghouse had adopted 60 Hz for lighting and 30 Hz for power and refused to guarantee efficiency at less than 30 Hz. General Electric recommended 41 2/3 Hz. Following a dinner meeting in October in New York City with Westinghouse representatives, President Adams of the Cataract Company asked Westinghouse's chief electrical engineer if they could build and guarantee a 25-Hz generator.

Later in October revised proposals were received from Westinghouse and General Electric. Changes had to be made in the Forbes design. A 20,000-volt oil-cooled armature was not practical.[66] On October 26 a contract was placed with Westinghouse for three 5000-hp, 25-Hz, 2200 volt, two-phase, four-wire generators.[67] Figure 5.10 shows a view of a generator armature at the Westinghouse factory. The armature conductors were insulated with mica, which was a fortunate choice because thermocouple tests in later years showed temperatures as high as 225C. Figure 5.9 shows a nickel steel field ring with two of the twelve poles and coils. Adams 'Niagara Power' book contains many pages detailing the design, manufacture, testing and installation of these pioneering generators along with the auxiliary electrical apparatus including exciters, measuring instruments and switching devices [Fig 5.11]. The testing of the first machine at the Westinghouse plant in Pittsburgh was probably the origin of the cardinal rule "Never place a short circuit on a piece of rotating machinery." This was done during testing with great disturbance to the windings.[68]

Fig. w

The first turbo-generator was successfully tested and operated in April 1895 [Fig 12].[69] The tall gentleman, No. 7, is John Jacob Astor and No. 8 is Edward Dean Adams, President. Schallenberger of Westinghouse designed a new type of switchboard indicating and integrating meters. On August 26, two-phase four-wire 2200 volts was first delivered commercially one-half mile to the Pittsburgh Reduction Company for reducing aluminum ore using the Hall process.[70] This company became ALCOA, the Aluminum Company of America. The Carborundum Company, maker of silicon carbide abrasives, was another early customer. The Forbes Subway, that served these early customers, was a concrete tunnel with cast iron racks on both walls for supporting the four copper conductor, rubber insulated, lead-covered cables in each circuit [Fig 13].[71]

The use of low cost Niagara power enabled these companies, and other soon to follow metallurgical and chemical companies, to greatly reduce the cost of their products. Aluminum is a classic example. In the mid 1800's Napoleon III had a set of aluminum forks and spoons made for his most honored guests; less important guests used gold or silver tableware.[72] In 1884 the cap for the Washington Monument was a 6¼-pound piece of aluminum, the largest made to that date in the United States. During 1905 and 1906 the Niagara, Lockport and Ontario Power Company used about 260 tons of aluminum overhead conductors to transmit Niagara power to Syracuse [Fig. 14].[73] By October 1896, when the third 5000-hp turbo-generator was placed in operation, the demand for electricity locally exceeded the capacity.[74] Arrangements were made for additional capacity.[75]

Fig. x

Resources

https://ericpdollard.com

For more books and videos by Eric Dollard, visit

https://emediapress.com/epd

Follow Eric on social media

https://ericpdollard.com/facebook

https://ericpdollard.com/twitter

Watch Eric's free videos and interviews

https://ericpdollard.com/youtube

Eric Dollard's Official Forum

https://ericpdollard.com/forum

http://www.energeticforum.com

https://www.energyscienceforum.com

https://emediapress.com

Become an affiliate and earn up to 45% commission

https://emediapress.com/affiliate-area

Meet Eric Dollard and other pioneers of the modern-day Tesla movement at the Energy Science & Technology Conference

https://energyscienceconference.com

Videos of previous conference presentations are available at

https://emediapress.com/estc

Lakhovsky Multiwave Oscillator and Bedini RPX Sideband Generator

https://vril.io

www.teslascientific.com

www.griffingbrock.com

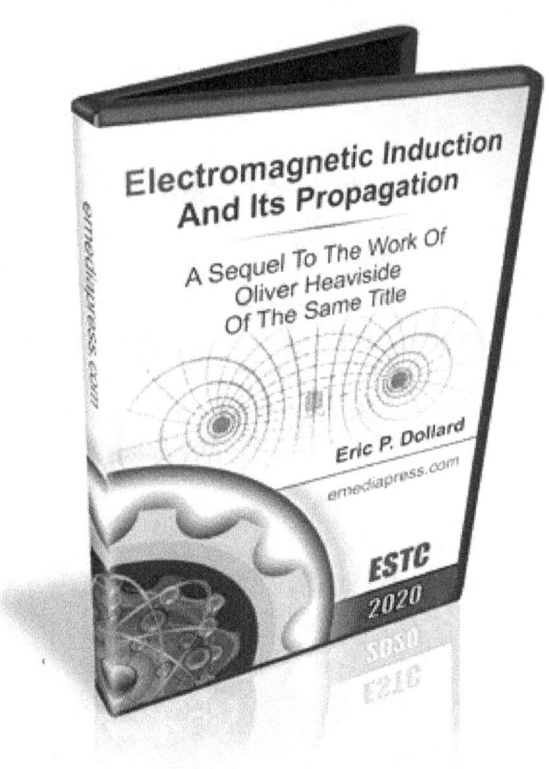

https://emediapress.com/emivideo

SUPPORT
EPD LABORATORIES, INC.
501(C)3 NON-PROFIT

Please make tax-deductible donations at:

https://ericpdollard.com/donate

Make Checks or Money Orders payable to:
EPD Laboratories, Inc.

EPD Laboratories, Inc.
PO Box 10029
Spokane, WA 99209

DONATE BITCOIN

BTC ADDRESS

1Cro4RfF4pFXwMa8pPVHXTk425YyXbufrW